ADVANCED TOPICS
IN SCIENCE AND TECHNOLOGY IN CHINA

ADVANCED TOPICS
IN SCIENCE AND TECHNOLOGY IN CHINA

Zhejiang University is one of the leading universities in China. In Advanced Topics in Science and Technology in China, Zhejiang University Press and Springer jointly publish monographs by Chinese scholars and professors, as well as invited authors and editors from abroad who are outstanding experts and scholars in their fields. This series will be of interest to researchers, lecturers, and graduate students alike.

Advanced Topics in Science and Technology in China aims to present the latest and most cutting-edge theories, techniques, and methodologies in various research areas in China. It covers all disciplines in the fields of natural science and technology, including but not limited to, computer science, materials science, life sciences, engineering, environmental sciences, mathematics, and physics.

Feng Xia
Youxian Sun

Control and Scheduling Codesign

Flexible Resource Management in Real-Time Control Systems

With 118 figures

ZHEJIANG UNIVERSITY PRESS
浙江大学出版社

Springer

AUTHORS:

Dr. Feng Xia
State Key Laboratory of Industrial Control Technology
Zhejiang University
310027, Hangzhou, China
E-mail: f.xia@ieee.org

Prof. Youxian Sun
State Key Laboratory of Industrial Control Technology
Zhejiang University
310027, Hangzhou, China
E-mail: yxsun@iipc.zju.edu.cn

ISBN 978-7-308-05765-3 **Zhejiang University Press, Hangzhou**
ISBN 978-3-540-78254-4 **Springer Berlin Heidelberg New York**
e-ISBN 978-3-540-78255-1 **Springer Berlin Heidelberg New York**

Series ISSN 1995-6819 Advanced topics in science and technology in China
Series e-ISSN 1995-6827 Advanced topics in science and technology in China

Library of Congress Control Number: 2008925538

Co-published by Zhejiang University Press, Hangzhou and Springer-Verlag GmbH Berlin Heidelberg

Springer is a part of Springer Science + Business Media
springer.com

Cover design: Joe Piliero, Springer Science + Business Media LLC, New York
Printed on acid-free paper

Preface

Recent evolutionary advances in information and communication technologies give rise to a new environment for Real-Time Control Systems. This is a new dynamic environment that features both resource limitation and workload variability. As a consequence, the availability of the computing and/or communication resources becomes typically uncertain in modern Real-Time Control Systems. In this context, the expected *Quality of Control* (QoC) of the systems cannot always be guaranteed by the traditional control systems design methodology that separates control from scheduling. From a resource scheduling perspective, the prevalent open-loop scheduling schemes in real-time systems obviously lack flexibility when applied to Real-Time Control Systems operating in dynamic environments. To make the best use of available resources, more holistic principles and methods need to be developed. These requirements motivate the recent technological trend towards the convergence of computing, communication and control.

This book is a monograph that covers our recent and original results in this direction. The main objectives of this work are:

(1) *To construct a unified framework of feedback scheduling that enables the integration of control with computing and communication.* This framework will encompass a set of concrete feedback scheduling methods and algorithms that are applicable to different systems. With these methods and algorithms, solutions are provided for some key issues in feedback scheduling, thus promoting the emergence of this area.

(2) *To enable flexible QoC management in dynamic environments with uncertainty in resource availability.* A number of new approaches to flexible management of the computing and/or communication resources in Real-Time Control Systems will be developed to maximize or improve the overall system performance.

With these objectives in mind, we focus on feedback scheduling strategies for flexible resource management in the context of real-time control. The traditional control systems design methodology and the simple control task model based on the fixed timing constraints are discarded. closed loop dynamic resource management schemes are built by means of control and scheduling codesign. The major tool used in this book is feedback scheduling. The introduction of feedback into dynamic resource management breaks the traditional open-loop mode of resource scheduling.

We are not interested in solutions that belong to any particular discipline, i.e. control theory, computer science or communication technology. Accordingly, we do not attempt to:

(1)*Design new control algorithms.* No innovation is made in this book regarding controller design that could make the control loops robust against delay, data loss or jitter. This is often difficult because it requires a solid foundation of mathematics. Furthermore, there has been extensive interest in this direction for many years, with an abundance of theoretical results produced.

(2)*Develop new real-time task scheduling policies.* Implementing a new scheduling policy demands support from the underlying platforms, e.g. the operating systems. Therefore, it is generally hard for a new scheduling policy to become practical even if it indeed performs better than existing ones in some situations. On the other hand, many mature scheduling policies are available in the area of real-time scheduling.

(3)*Develop new communication protocols.* Despite rapid evolution, none of the wireless technologies in existence was designed particularly for control applications. Intuitively, developing such a dedicated communication protocol (from scratch or based on some existing protocols) could better support wireless control. However, this is beyond the scope of this book due to its emphasis on networking.

While the framework presented is applicable to a broad variety of computing and communication resources, special attention of this book is paid to three representative classes of resources, i.e. CPU time, energy and network bandwidth. By exploiting the feedback scheduling methodology, we develop a set of resource management schemes. Numerous examples and case studies are given to illustrate the applicability of these schemes.

The inherent multidisciplinary property of the codesign framework makes the intended audience for this book quite broad. The first audience consists of researchers interested in the integration of computer science, communication technology and control theory. This book presents a unified framework as an enabling technology for codesign of computing, communication and control. Novel paradigms for Real-Time Control Systems research and development in the new technological environment also provide insight into new research directions in this emerging area.

The second audience is practitioners in control systems engineering as well as computer and communication engineering. Careful balance between theoretical foundation and real-world applicability makes the book a useful reference not only for academic research and study but also for engineering practice. Much effort has been devoted to make this book practical. For instance, the problems addressed are of remarkably practical importance; all solutions are developed with the practicability in mind.

This book is also of value to graduate students in related fields, for whom the tutorial introduction to feedback scheduling and the extensive references to the literature will be particularly interesting. The background of the reader assumed in the presentation encompasses a basic knowledge in feedback control theory, sampled-data control, and real-time systems. Prior experience with intelligent control, power-aware computing, and network bandwidth allocation is also helpful, though not necessary.

Outline of the Book

This book is broken into four parts, each part containing two chapters. The first part, Background, covers Chapters 1 and 2, in which the motivations, background information and basic framework for this work are given. Part II (CPU Scheduling) is concerned with scheduling the shared processor in multitasking embedded control systems. Chapters 3 and 4 belong to this part. Their focuses are on developing intelligent feedback scheduling methods by exploiting neural networks and fuzzy control respectively. While CPU scheduling is also involved, the main concern of Part III (Energy Management) that covers Chapters 5 and 6 is dynamic management of the energy consumption of embedded controllers. The general goal of this part is to reduce the CPU energy consumption while preserving QoC guarantees. The last part, Bandwidth Allocation, which covers Chapters 7 and 8, studies how to dynamically allocate available communication resources among multiple loops in networked control systems and wireless control systems respectively.

In Chapter 1 we give an overview of the field of control and scheduling codesign. The motivations for codesign of control and scheduling are illustrated. Recent trend towards the convergence of computing, communication and control is pointed out. After related work in the literature is reviewed, a perspective on feedback scheduling of Real-Time Control Systems is given. Chapter 2 presents a tutorial introduction to feedback scheduling. Background knowledge about sampled-data control and scheduling in real-time systems is outlined, which makes this book more self-contained. The temporal attributes of control loops are described. Motivating examples for applying feedback scheduling are presented. Key concepts, basic framework, and design considerations related to feedback scheduling are also described.

Chapter 3 concerns neural feedback scheduling. The primary goal is to overcome the disadvantages of overly large computational overhead associated with mathematical optimization routines. We present a fast feedback scheduling scheme to exploit feedforward neural networks. This scheme can dramatically reduce feedback scheduling overhead while delivering almost optimal overall control performance.

Chapter 4 presents a fuzzy feedback scheduling scheme based on fuzzy logic control. This scheme is independent of task execution times, robust to measurement noise, while handling uncertainty in resource availability in an intelligent fashion. Being easy to implement, the fuzzy feedback scheduler also incorporates quite a small runtime overhead.

Considering the unpredictability of task execution times as well as the variability of CPU workload, Chapter 5 develops a feedback control real-time scheduling methodology called Energy-Aware Feedback Scheduling. It integrates the management of both energy consumption and QoC. After analytically modelling the Dynamic Voltage Scaling system, a control-theoretic design and analysis method for feedback schedulers is proposed. Taking advantage of the basic framework of Energy-Aware Feedback Scheduling. Chapter 6 aims to achieve further energy consumption reduction while not jeopardizing the quality of control. For this purpose, we present an Enhanced Energy-Aware Feedback Scheduling scheme to exploit the methodology of graceful gradation.

Chapter 7 pays attention to multi-loop networked control systems. By exploiting codesign of control and network scheduling, we develop an integrated feedback scheduling scheme. This scheme can maximize resource utilization in the case of light workload and achieve graceful performance degradation under overload conditions. To attack the uncertainty in available communication resource in wireless control systems, Chapter 8 presents a cross-layer Adaptive Feedback Scheduling scheme that takes advantage of cross-layer design. It proves quite efficient to deal with channel capacity variations and noise interference. We also suggest an event-driven invocation mechanism for feedback schedulers to further improve the feedback scheduling performance.

We consider only linear time-invariant (LTI) systems in this book, though the proposed approaches are not only applicable to this class of systems. The control applications used as simulation examples are kept typical and illustrative to exhibit the wide applicability of the proposed approaches. For instance, the controlled processes studied include inverted pendulum, DC motor, and many others. The control algorithms cover PID (Proportional-Integral-Derivative), state feedback control with pole placement, LQG (Linear Quadratic Gaussian) controllers, etc. The control theory involved comprises classical control, modern control, as well as intelligent control. In addition, both of the two design methods for digital controllers i.e. discrete-time design and discretization of continuous-time controllers have been adopted in different simulations. Throughout the book, all simulations are conducted using Matlab/Simulink[1] along with the TrueTime[2] toolbox.

---

1 Matlab and Simulink are registered trademarks of The MathWorks, Inc.

2 http://www.control.lth.se/truetime/

Acknowledgements

This book summarizes our research results achieved in recent years. Many people contributed to this book in various ways. We would especially like to thank Yu-Chu Tian at Queensland University of Technology, Moses O. Tadé at Curtin University of Technology, Chen Peng at Nanjing Normal University, Wenhong Zhao at Zhejiang University of Technology, Jinxiang Dong, Zhi Wang and Xiaohua Dai at Zhejiang University, Liping Liu at Tianjin University, for their collaboration. We are grateful to Zonghai Sun at South China University of Technology, Xiaodong Wang, Jianhui Zhang and Peng Cheng at Zhejiang University who reviewed earlier drafts. Special thanks go to Li Yu at Zhejiang University of Technology and Russell Morgan at Imprimis Computers, Brisbane, who read the manuscript line by line and gave particularly helpful suggestions for improvements, as well as Qing Lin at Zhejiang University, who helped set up wonderful working environments.

We gratefully acknowledge all support from Springer and Zhejiang University Press.

We dedicate this book to our families.

Feng Xia
Youxian Sun

Notations and Symbols

AFS	Adaptive Feedback Scheduling
α	Voltage scaling factor, normalized CPU speed
$\alpha_{\min}$	Minimum allowable voltage scaling factor
BP	Back Propagation
c_i	Execution time of task i at actual CPU speed
$c_{i,\mathrm{nom}}$	Execution time of task i at full CPU speed
$\hat{c}_i$	Estimated execution time of task i at actual CPU speed
$\hat{c}_{i,\mathrm{nom}}$	Estimated execution time of task i at full CPU speed
COTS	Commercial Off-The-Shelf
CPU	Central Processing Unit
DVS	Dynamic Voltage Scaling
E	Normalized energy consumption of CPU
EAFS	Energy-Aware Feedback Scheduling
ECS	Embedded Control System
EDF	Earliest Deadline First
EEAFS	Enhanced Energy-Aware Feedback Scheduling
FFS	Fuzzy Feedback Scheduling
h_i	Sampling period of control loop i, period of task i
$h_{i,0}$	Nominal (initial) sampling period of task i
$h_{i,\max}$	Maximum allowable sampling period of control loop i
$h_{\min}$	Minimum allowable sampling period
IAE	Integral of Absolute Error
IFS	Integrated Feedback Scheduling
j	Invocation instant of a feedback scheduler
J_i	Control cost of control loop i
J_{SUM}	Total control cost of all control loops
k	Sampling instant of a control loop

K_D	Derivative coefficient of a PID controller
K_I	Integral coefficient of a PID controller
K_P	Proportional coefficient of a PID controller
MAC	Medium Access Control
N	Number of concurrent control loops/tasks
NCS	Networked Control System
NFS	Neural Feedback Scheduling
OFS	Optimal Feedback Scheduling
OLS	Open-Loop Scheduling
opDVS	optimal pure Dynamic Voltage Scaling
OSI	Open System Interconnection
PID	Proportional-Integral-Derivative
ρ	Deadline miss ratio
ρ_r	Setpoint for deadline miss ratio
QoC	Quality of Control
QoS	Quality of Service
r	Transmission rate of a network
r_i	Reference input of control loop i
RM	Rate Monotonic
RTCS	Real-Time Control System
SQP	Sequential Quadratic Programming
T_{FS}	Invocation interval of a feedback scheduler
U	CPU/Network utilization
U_{req}	Requested utilization
U_R	Utilization setpoint
w_i	Weighting coefficient of control loop i
WCET	Worst-Case Execution Time
WCS	Wireless Control System
WLAN	Wireless Local Area Network
y_i	System output of control loop i
η	Period scaling factor

Contents

PART I BACKGROUND

PART II CPU SCHEDULING

PART III ENERGY MANAGEMENT

PART IV BANDWIDTH ALLOCATION

PART I
BACKGROUND

Chapter 1

Overview

This chapter gives an overview of the field of control and scheduling codesign, thus constituting the background of this work. The features of modern real-time control systems are examined with respect to resource availability. The potential disadvantages of the traditional control systems design and implementation method when used in environments featuring resource limitation and workload variations are characterized. Thus the motivations for codesign of control and scheduling are illustrated. Recent trend towards the convergence of computing, communication and control is pointed out. With emphasis on feedback scheduling and real-time control, related work in the two research directions, integrated control and computing and integrated control and communication, is roughly reviewed respectively. A perspective on feedback scheduling of Real-Time Control Systems is also given.

1.1 Real-Time Control Systems

In recent years there are revolutionary changes in information and communication technologies [MUR03]. With the availability of ever more powerful and cheaper products, the roles that embedded computing and networked communication play in control systems engineering are becoming increasingly important. The number of networked embedded devices deployed in practice has been far more than that of various general-purpose computers including, e.g. desktop PCs [ARZ05a, BOU05, LOY03, XIA06b]. These devices are used in many application areas such as aerospace, instrument, process control, communication, military, consumer electronics, etc. In this context, networked embedded control systems have become unprecedentedly popular.

There is no doubt that control systems constitute an important subclass of real-time systems in which the value of the task depends not only on the correctness of the computation and/or communication but also on the time at which the results are available [LIU00, STA96]. Since almost all control systems in the real world are implemented upon specific computing and communication platforms using digital technologies, the temporal behaviour is fundamentally crucial to the design and implementation of control systems.

From a point of view of real-time systems, the temporal behaviour of a system highly relies on the availability of resources. It is compulsory for the system to gain sufficient resources within a certain time interval in order that the execution of individual tasks can be completed in time. In a typical modern Real-Time Control System (RTCS) there are mainly two categories of resources, i.e. computing resources and communication resources. The former includes for example the processing time of the processor (i. e. CPU), which is usually called the controller by control engineers, and the energy budget of the battery, while the bandwidth of the communication network is a commonplace example of the latter.

It has been recognized that most embedded implementation platforms for Real-Time Control Systems are suffering from resource limitations [ARZ99b, ARZ03, PAL02a, XIA04a, XIA04b], which is in contrast to general-purpose computing and communication systems. The reasons behind are manifold. Technically speaking, embedded devices are often subject to various limitations on physical factors such as size and weight due to the stringent application requirements. A direct consequence is the limitation on the processing power, i.e. the allowable speed of the processor. To guarantee determinism in real-time communications, it is only possible for control networks to support data transmission rates that are dramatically lower than what non-real-time networks, e.g. Ethernet, can offer [BEN06, THO05]. For an increasing number of embedded platforms that run on energy supplied by batteries, the amount of energy available for the system has become a major constraint. This is certainly the truth today because the progress of battery technology cannot keep pace with the increase in energy expenditure [UNS03].

In order to cut down the cost, in most practical cases the commercially available processors with the highest allowable speeds and networks with the largest allowable bandwidths will seldom be deployed. This is mainly because they are always accompanied by excessively high costs. Therefore, the computing and communication platforms used in practice are usually equipped with limited resources that are considered to be sufficient for satisfying the requirements of the target applications. This is of course another important reason for the resource limitations associated with modern Real-Time Control Systems.

In spite of resource limitations, the complexity of real-world control applications is continuously growing [ARZ05a, MUR03]. In many situations, a lot of factors including the increase in functionality requirements from users and the permanently competitive market are driving the systems to become more complex than ever. As a consequence, the dedicated processors traditionally available for individual control tasks in embedded systems will nowadays cease to exist in most of Real-Time Control Systems. Instead, different tasks have to compete for the use of the same processor on which they run concurrently. In a sense, the presence of multiple tasks could potentially add to the burden of the processor with respect to energy consumption. As the systems become more complex, the traditional point-to-point communication mode has been replaced by the networked architecture where diverse communication entities need to transmit data over the same network. Intuitively, this sharing of the system resource with other tasks makes the limitations of resources even more pronounced for individual tasks, which could substantially affect the temporal behaviour of the corresponding control systems.

It is not uncommon today that many Real-Time Control Systems have to operate in dynamic, uncertain environments that feature workload variations [BUT04, XIA06b]. To meet stringent requirements of the ever changing market, the flexibility, reconfigurability, expandability, etc. of engineering systems have been enhanced remarkably. These features make it possible to reconfigure the systems during runtime, thus easing system maintenance and update. However, a natural result of online reconfiguration is that the system workload will change accordingly. For instance, some functional modules within the system may need to be removed in some cases, while new modules may be added in other cases. Some applications' demands on the shared resources could even be changed simply because of the update of the relevant functional modules. In these circumstances, the system workload will vary, which in turn influences the availability of the shared resources for control tasks, no matter whether or not the control modules are involved in the system update. With resource limitations, this variability of workload inevitably leads to the fact that the uncertainty in resource availability has become a critical bottleneck challenging the design and implementation of Real-Time Control Systems.

It is unfortunate that the uncertainty in resource availability has been accentuated by the application of Commercial Off-The-Shelf (COTS) components to Real-Time Control Systems [ARZ05a, MUR03]. There have been a large number of attempts to apply non-real-time operating systems such as Linux, Windows CE, and TinyOS as well as non-real-time communication networks including Ethernet [YIN04] and WLAN (Wireless Local Area Network) to various control systems. The reasons behind these applications mainly relate to the cost and/or the average system performance. Due to the adverse properties of these COTS components, however, temporal determinism cannot be supported.

From a control point of view, the uncertainty of resource availability incurs temporal non-determinism among control loops, which is often reflected by time-varying delays and data losses. As a result, the *Quality of Control* (QoC) of the system may be deteriorated, and in extreme cases the stability may be jeopardized. The notion of QoC [MAR02a, SAN04] used here originates from the combination of the *Quality of Service* (QoS) concept from the computer and communication communities and the control performance concept from the control communities, thereby representing a result of the convergence of computing, communication and control.

Demerits of Control and Scheduling Separation

Traditionally, the engineering process of a control system consists of two steps: controller design and its implementation, which are always separated [ARZ99b,

MAR02a]. While the controller design is usually done by control engineers, the implementation is the responsibility of system (or computer) engineers. In this way, the control engineers pay no attention to how the well-designed control algorithms will be implemented, while the system engineers are not aware of the requirements of the control applications with respect to temporal attributes. The most notable reason for this phenomenon is that control and real-time scheduling are two disciplines that have been evolving separately for the past many years [MUR03]. By default, control engineers often assume that:

(1) the implementation platforms can support equidistant and periodic sampling,
(2) the computational and communication delays are negligible or constant, and
(3) the data will never be lost.

In like manner, the system engineers always assume that the control tasks:

(1) are periodic with a set of fixed periods,
(2) have hard deadlines, and
(3) have known Worst-Case Execution Time (WCET).

In fact, all of these assumptions are not always real, especially for systems operating in dynamic environments with uncertainty [ARZ99b]. In practical embedded control systems, for instance, a large number of control tasks are characterized by execution times that are not constant due to data dependency or varying application requirements. Instead, their execution times may significantly change from instance to instance, which implies a large delay jitter. It is also the case that physical platforms cannot provide all control loops with equidistant sampling because of the presence of resource contention by multiple applications. Therefore, sampling (period) jitter is unavoidable. In real networked control systems delay jitter and sampling jitter are also present. Furthermore, the use of communication networks (especially wireless networks) makes it more likely that the transmitted data be lost.

While control systems use fixed sampling periods in most cases, the simple task model with fixed periods will not be applicable to Real-Time Control Systems that employ event-driven control [HEN06] or feedback scheduling [ARZ05b, ARZ06a, SHA04], as well as those control systems that are characterized in nature by aperiodic sampling, for example, the fuel injection control system for an automotive engine. Hard real-time constraints are overly rigid for a variety of control applications. For most control systems, occasional deadline misses do not necessarily yield catastrophic results, since real systems are always designed with sufficient robustness and stability margin. Consequently, a large number of control systems are not hard real-time but soft real-time. There are also many control tasks whose WCET are not available because of the inherent complexity of the control algorithms they execute. Even if the WCET are known, the traditional design method based on WCET cannot make full and reasonable use of available resources.

From a view of resource scheduling, classical priority-driven scheduling policies such as RM (Rate Monotonic) and EDF (Earliest Deadline First) [LIU73] are based on complete and accurate *a priori* knowledge about timing attributes of real-time tasks. They can maximize the temporal determinism under deterministic and resource-sufficient environments. However, these scheduling policies do not take into account the requirements of target applications that originate from their actual performances. When applied to Real-Time Control Systems with limited resources and variable workloads, they cannot maintain the system schedulability under overload conditions. When the system is underloaded, some resources will be wasted, thus causing the control performance to be worse than possible. From a feedback control point of view, these scheduling policies are open-loop [LU02a]. Once configured at design time, they do not intentionally change any scheduling parameters during runtime, e.g. according to some feedback information. As a consequence, traditional scheduling policies are far from flexible when used for scheduling the uncertain, limited resources in multi-loop Real-Time Control Systems. In dynamic environments, the required quality of control cannot always be guaranteed with these scheduling policies.

To summarize, the traditional control systems design and implementation method features separation of control and scheduling. It cannot meet the requirements of real-time control in the new environment with uncertainty in resource availability. Therefore, it is necessary to develop a new paradigm for resource management in Real-Time Control Systems. It could be envisioned that such a paradigm would take advantage of the interplay between computing, communication and control. Actually, this is currently one of the dominant technological trends in information and communication technologies.

1.2 Convergence of Computing, Communication and Control

A lot of challenges are facing the field of feedback control, especially when the new environment that differs substantially from that of the past half century is rapidly emerging [MUR03]. Regardless, the rapid advances in computing, communication, sensing and networking also give rise to unprecedented opportunities for Real-Time Control Systems, as exhibited by the following summary from the NSF/CSS Workshop on New Directions in Control Engineering Education [ANT99, MUR03]:

> *The field of control systems science and engineering is entering a golden age of unprecedented growth and opportunity that will likely dwarf the advancements stimulated by the space program of the 1960s. These opportunities for growth are being spurred by enormous advances in computer technology, material science, sensor and actuator technology, as well as in the theoretical foundations of dynamical systems and control. Many of the opportunities for future growth are at the boundaries of traditional disciplines, particularly at the boundary of computer science with other engineering disciplines. Control systems technology is the cornerstone of the new automation revolution occurring in such diverse areas as household appliances, consumer electronics, automotive and aerospace systems, manufacturing systems...*

As mentioned above, in the new environment characterized by widespread applications of networked embedded devices, the traditional standard design pattern, which allows the separation of different disciplines, will on longer be valid when applied to Real-Time Control Systems engineering. To address this problem, principles and methods in the computer and communication communities should be incorporated into control systems design and implementation. The feedback control theory and technology could of course also be adopted in both computing and communication systems to achieve better performance. This interplay is leading to the emerging wave of information technology revolution, i.e. the convergence of computing, communication and control [GRA03, GRA04a, MUR03, XIA04b].

In the past half century, the fields of computer science, communication and control have achieved great successes in isolation. The accomplishment of ENIAC, the world's first electronic, large-scale, general-purpose computer, activated at the University of Pennsylvania in 1946 is a symbol of the birth and evolution of modern computer science. The paper entitled "A mathematical theory of communication" by Shannon, published in 1948 [SHA48], builds the foundation of information theory. The publication of the book "Cybernetics" by Wiener in 1948 [WIE48] symbolizes the naissance of control science, a discipline of which the principles and methods have been widely used today in various engineering systems. The rich theoretical and technological results achieved in these disciplines have formed a solid foundation for the development of a new unified system theory as a result of the convergence. For instance, the recent rapid progress in pervasive computing, communication and sensing technologies brings new opportunities to real-time control. Embedded processors and wireless networks will certainly become the characteristic components of future Real-Time Control Systems [MUR03].

As a matter of fact, the convergence of computer science and communication has given birth to a series of important technologies, of which the most representative examples include the Internet and wireless technology. More exciting is that there has been a great deal of evidence of the trend towards the integration of computing, communication and control.

For instance, there are an increasing number of researchers from the computer and communication communities attending some classical conferences in control, e.g. the IEEE Conference on Decision and Control (CDC) and the American Control Conference (ACC). On the other hand, more and more researchers from the control community are interested in attending a variety of conferences in computer science and communication, e.g. the IEEE Real-Time Systems Symposium (RTSS) and the IEEE Conference on Computer Communications (INFOCOM). In addition, quite a few academic conferences and journals covering multiple disciplines such as computer science, communication and control have been launched in recent years. These phenomena disclose that the traditional strict boundaries between these disciplines will become less and less rigorous. Instead, the dynamic interaction between these disciplines will be a key feature of further information technology.

To promote the emergence of this multidisciplinary area, the Panel on Future

Directions in Control, Dynamics, and Systems have recommended that the government agencies and the control community should substantially increase research aimed at the integration of control, computer science, communications, and networking [ANT99, MUR03]. A number of institutions have started related research initiatives. For example, the U.S. Department of Defense has already made a large investment in this area through the multidisciplinary university research initiative (MURI) program. It has been reported that the U. S. Department of Defense planned to continue to award research grants with a total of over $150 million to academic institutions under this program over the five years from 2006 [DOD06]. Table 1.1 gives some other examples of research initiatives related to the convergence of computing, communication and control.

Table 1.1 Examples of research initiatives related to convergence of computing, communication and control

Program	Objectives	Sponsors	Institutions	Term
Q-MARS[1]	To provide QoS support for surveillance and control systems, including QoS-based resource allocation, quantized EDF scheduling, etc.	US: DARPA, ONR	CMU, UIUC, UVa	2001 –
MVWT[2]	To develop a tool for validating theoretical advances in multiple-vehicle coordination and control, networked control, real-time networking and high-confidence distributed computation	US: DURIP, AFOSR	Caltech	2001 – 2003
FLEXCON[3]	To provide design and implementation techniques for embedded control systems that support runtime flexibility with respect to changes in, e.g. workload and resource utilization patterns	Sweden: SSF	LTH, KTH, ABB, etc.	2003 – 2005
ARTIST2[4]	To build a durable European research community on embedded systems design by integrating 7 clusters: modelling and components, hard real-time, adaptive real-time, compilers and timing analysis, execution platforms, control for embedded systems, testing and verification	Europe: FP6 IST	Multiple	2004 – 2008
RUNES[5]	To enable the creation of large-scale, widely distributed, heterogeneous networked embedded systems that adapt to their environments	Europe: FP6 IST	Multiple	2004 – 2007
WiSA[6]	To develop networking protocols, sensor fusion techniques, and control methods that work in harmony, enabling wide deployment of wireless industrial automation and monitoring applications	Finland: Tekes, Sweden: Vinnova	UV, TKK, KTH	2006 – 2007

[1] http://www.cs.cmu.edu/~rtml/muri/; [2] http://www.cds.caltech.edu/~mvwt/; [3] http://www.control.lth.se/flexcon/; [4] http://www.artist-embedded.org/fp6/; [5] http://www.ist-runes.org/; [6] http://tite.uwasa.fi/wisa/

Besides these projects listed in Table 1.1, there are of course many others in related directions. Regardless of the difference in the main topics, a common objective of these programs is to promote the systematic convergence of multiple disciplines in terms of both theory and technology. In the context of real-time control, there are intuitively two kinds of convergences: integrated control and computing, and integrated control and communication. This book is concerned with these two aspects, without investigating the interplay between computer science and communication.

As shown in Fig.1.1, the integration of computing, communication and control offers a fresh methodology for implementing real-time control in dynamic environments. Following this methodology, it is thus possible to realize the codesign of computing, communication and control in control systems engineering, which is in contrast to the traditional design pattern that separates control and scheduling. The whole process of control systems implementation will never again be composed of these two separated stages. While the controller design will take into account the constraints of implementation platforms, the well-designed control algorithms will be implemented by the system engineers with the timing requirements of control applications in mind. In this way, a fundamental process of dynamic interaction is established between computer science, communication technology and control systems, which is believed to be able to contribute to system performance optimization [ARZ05a].

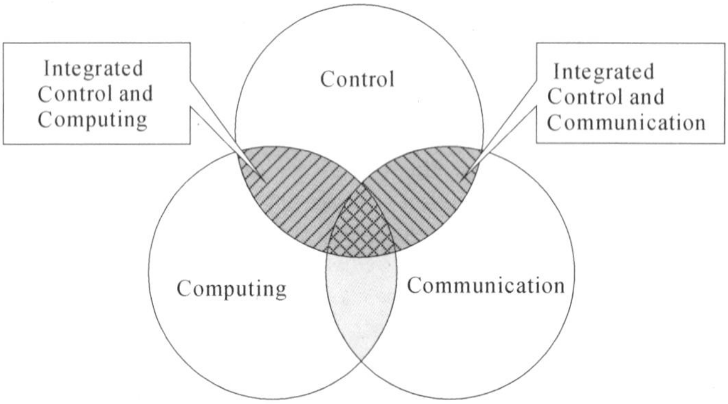

Fig. 1. 1 Schematic diagram for integration of computing, communication and control

However, it is still an open question how to build a more holistic theory that is essential for future progress in the convergence of computing, communication and control [GRA03, MUR03]. As an initial effort in this direction, this book focuses on the flexible management of uncertain computing and/or communication resources in Real-Time Control Systems. Since resource scheduling becomes the main concern in this context, this area is often referred to as *control and scheduling codesign* or *integrated control and scheduling*, which covers both areas of *integrated control and computing* and *integrated control and communication*.

In the last decade, there has been increasing interest in codesign of control and scheduling. Some of the efforts will be briefly outlined with respect to integrated control and computing and integrated control and communication, respectively, in the subsequent two sections. The emphasis is laid on: 1) the application of feedback scheduling to computing and communication systems, and 2) codesign of control and scheduling in Real-Time Control Systems. However, no claim is made that an exhaustive survey of all state-of-the-art work will be produced, which is of course impossible. The interested reader is referred to related journals and conference proceedings (e.g. RTSS, INFOCOM, CDC, ACC, etc.) for more recent work. For basic concepts of feedback scheduling, see Chapter 2.

1.3 Integrated Control and Computing

The majority of existing work in the area of integrated control and computing can be divided into two categories: control of computing systems and computer based control. In the former, the applications considered are usually general-purpose computing systems (i.e. non-control systems), and the key technology used for resource management is feedback control/scheduling. On the contrary, the latter mainly considers embedded control systems, a subclass of Real-Time Control Systems, wherein the control performance is a natural concern.

1.3.1 Control of Computing Systems

As the name tells, the basic idea behind control of computing systems is applying feedback control theory and techniques to computing systems such that some desired performance is achieved. Feedback control provides the engineering community with an enabling technology to attack uncertainty and improve flexibility of the systems [HEL04]. As a well-established discipline that has been successfully applied to dynamic adjustment of physical systems, feedback control brings obvious advantages to computing systems [ARZ06a]. For instance, when feedback control is introduced, computing systems' robustness to external and/or internal disturbances can be significantly enhanced, thus improving the system determinism during runtime; feedback control can also be used to compensate for or reduce the negative effects of implementation platforms on system performance. In particular, for computing systems operating in dynamic open environments, using feedback control in resource management could improve the system flexibility in the face of uncertainty in resource availability, which is of paramount importance for guaranteeing satisfactory temporal behaviour of the systems.

Computing systems are certainly a new application area for feedback control theory. Due to the inherent complexity of computing systems, in most cases it is hard to construct (sufficiently accurate) mathematical models to formulate their dynamics. Sometimes even selecting an appropriate model type is extremely difficult. Therefore, it is not easy even for a skilled control engineer to realize effective control of computing systems [ARZ06a, HEL04, MUR03]. For this reason, it is not until recently that feedback control theory and techniques have been substantially applied to computing systems engineering.

In general, the essential of control of computing systems is to adjust dynamically the scheduling of some shared resources. Consequently, these control methods are often called *feedback control (real-time) scheduling* [ARZ06a, LU02b], from a resource management point of view. By defining some relevant variables including e.g. the controlled variable and the manipulated variable, the problem of dynamic resource management is formulated as a feedback control problem, thus introducing a feedback control loop into the computing systems. The feedback control scheduler can be designed based on feedback control theory by treating the computing systems as a controlled process. In contrast to traditional resource scheduling methods, which are often based on offline analysis, the feedback control scheduler gathers feedback information about actual resource usage at runtime and compares it with the desired level. According to their difference, the current available resource will be re-allocated online among different users by exploiting some control algorithms.

CPU Resource Control

As one of the first work in this direction, Stankovic, *et al.* [LU99, STA99] present a scheduling framework called FC-EDF (Feedback Control EDF) taking advantage of feedback control theory and methods. In this framework, the feedback scheduler monitors the task deadline miss ratio and/or CPU utilization at runtime. A PID (Proportional-Integral-Derivative) controller is used to adjust the CPU utilization estimate of a set of soft real-time tasks with varying execution times via a QoS actuator. The objective is to achieve both lower deadline miss ratio and higher CPU utilization. The CPU consumption of the tasks can be changed by executing different versions of the same algorithm. In addition, larger changes in the workload are dealt with using an admission controller. In [LU00] and [LU02a] the approach is extended and further generalized. In [STA01] Stankovic, *et al.* propose a framework called distributed feedback control real-time scheduling (DFCS) for distributed real-time systems. The framework is based on feedback control that incrementally corrects system performance to achieve its target in the absence of initial load and resource assumptions.

Following this idea, Lu, *et al.* [LU05] present an end-to-end utilization control algorithm that adaptively maintains desired CPU utilization through performance feedback loops. The algorithm is based on a Model Predictive Control (MPC) approach that models utilization control on a distributed platform as a multivariable constrained optimization problem. A decentralized approach to end-to-end utilization control is presented in [WAN05] for large distributed real-time embedded systems. Based on hybrid supervisory control, Koutsoukos, *et al.* [KOU05] propose a feedback control scheduling method for real-time systems that operate in a finite set of timing and performance configurations.

Abeni, *et al.* [ABE00] propose a closed loop method based on adaptive control techniques for adapting online the fraction of assigned resource to the task requirements. In [ABE02, ABE05] the authors present an adaptive reservation method by applying a feedback scheme to a reservation-based scheduler. A mathe-matical model of the reservation-based scheduler is deduced. Also described therein is how this model can be used for synthesising the PI (Proportional-Integral) controller by applying results from control theory. Cucinotta, *et al.* [CUC04] implement the approach in a Linux environment.

In [PAL03] Palopoli, *et al.* model the reservation-based scheduler as a discrete-time switching system, and use hybrid control techniques to design the feedback scheduler. Song, *et al.* [SON05] propose to model the reservation-based real-time CPU scheduler as a feedback-controlled switched system with bounded time-varying uncertainty. A switched Feedback Bandwidth Server (sFBS) is presented for controller-based scheduling. The stability of the modelled system is analysed by the Lyapunov function. Using Linear Matrix Inequalities (LMI), a feedback control scheduler that stabilizes the switched system is designed. Altisen, *et al.* [ALT02] present a modelling methodology based on the controller synthesis paradigm, which allows a unified view of scheduling theory and approaches based on timing analysis of models of real-time applications.

In [SAH02] Sahoo, *et al.* present a closed loop approach for dynamically estimating the execution times of tasks based on both deadline miss ratio and task rejection ratio in the system. Therein an open-loop dynamic scheduling algorithm based on a notion of task overlap in the scheduler is exploited. Both a PI controller and a H_∞ controller are designed for closed loop scheduling. The performance of the approach is extensively evaluated by simulation and modelling in [ALO03]. Lin and Manimaran [LIN03b] design a double-loop feedback scheduler by integrating a local and a global feedback scheduler. Similar to [LU02a], the work aims to keep the deadline miss ratio near the desired level and to achieve high CPU utilization. In [LIN04a] Lin, *et al.* apply a feedback control scheduling approach to real-time systems with mobile nodes where the characteristics of mobility affect task parameters. In a concrete example of autonomous vehicle systems, low miss ratio and high CPU utilization are achieved.

Ravindran, *et al.* [RAV01] propose adaptive resource management techniques based on feedback control techniques for asynchronous, decentralized real-time systems. Different schedulers exploiting PID control and fuzzy logic control are designed respectively. Jin, *et al.* [JIN04] present a feedback control scheduling algorithm based on fuzzy control. The framework consists of a basic scheduler and a fuzzy feedback controller. The fuzzy scheduling algorithm is used as the scheduling policy in the basic scheduler. A fuzzy controller and a task flow adjustment policy constitute the fuzzy feedback controller.

Wei and Yu [WEI03] present a software real-time scheduling algorithm based on the hybrid adaptive feedback control architecture. Steere, *et al.* [GOE04, STE00] develop the real-rate scheduling method. The PID control algorithm is used to design a feedback-based adaptive scheduler. Tong, *et al.* [TON04] present a time-sharing scheduling algorithm based on PID feedback control. Lawrence, *et al.* [LAW01] present an adaptive CPU scheduling approach for dynamic real-time systems using control-theoretic methods. The approach employs the optimal feedback control to allocate CPU resource among periodic tasks in a way that a certain performance measure is optimized. Du and Levy [DU04] describe a H_∞ nonlinear model predictive control scheduling structure, with attempts to integrating it with the ORB middleware technology.

Zhang, *et al.* [ZHA04] present a feedback-driven approach for online scheduling of tasks with imprecise computation. The approach uses a PID controller to determine the percentage of the optional part to be admitted when a task comes into the system. Wei, *et al.* [WEI02] carry out a preliminary study about how to select the sampling periods for feedback control schedulers. A simple solution for varying the sampling period is presented. With the goal of improving the performance of feedback control scheduling, Amirijoo, *et al.* [AMI05] quantize the disturbance present in the measured variable as a function of the sampling period and propose a measurement disturbance suppressive control structure. Although feedback control theory and methods are not used explicitly, Buttazzo, *et al.* present an interesting method in [BUT02], where the feedback methodology and the elastic scheduling method are embedded. The task periods are adjusted according to the prediction of workload, so as to achieve the desired utilization level.

Server Systems Control

A great deal of research attention has been paid to control of server systems since the 1990s [HEL04, HEL05]. Here only those efforts related to feedback scheduling will be summarized briefly. In [HEL05] Hellerstein, *et al.* point out that control theory will play a fundamentally important role in the process of developing new server systems, particularly complex software systems. Therein they also discuss various challenges for applying control theory to server systems.

In server systems such as Web servers and email servers, there are generally three classes of control problems to be addressed [HEL04]. The first is about enforcing service level agreement in that customers receive the service levels for which they contracted, e.g. [LU01]. The second is to regulate resource utilization to avoid overload. For example, Sha, *et al.* [SHA02] apply PID feedback control to the Apache HTTP server based on a queuing model. The last is the optimization of system configuration. For instance, Dial, *et al.* [DIA02] propose to minimize the response time of the Apache Web server by exploiting fuzzy control.

A common approach to control of server systems is using the nonlinear model derived from queuing theory and a combined PID feedback control loop, a typical example of which is [SHA02]. Some server systems control methods based on advanced control theory have also been presented recently. For instance, Wei and Xu [WEI05] propose an adaptive fuzzy control approach to allocating resources in servers within a framework called eQoS. Xu, *et al.* present the dynamic resource allocation approach to exploiting predictive control in [XU06b], where different versions of the algorithm are compared. Liu, *et al.* [LIU06] propose a queuing model based adaptive control scheme for autonomous performance control of Web applications, where a queuing model predictor and an online adaptive feedback control loop are used to enforce admission control of the incoming requests, thus ensuring that the desired response time target is met.

Energy Management

In contrast to the above two directions, there is only a relatively small amount of progress that has been made in applying feedback control/scheduling to energy management. With regard to energy management, this book pays attention to only the Dynamic Voltage Scaling (DVS) [AYD04, KER05, PIL01, WEI94] technology due to its general popularity in this field. Described below are existing Dynamic Voltage Scaling methods for non-control systems. Energy management approaches for Real-Time Control Systems will be summarized later.

In [VAR03] Varma, *et al.* develop a PID-like algorithm, namely nqPID, to predict the workload for the DVS system, which delivers quite good performance. To minimize the energy expenditure of the processor while meeting the QoS requirements of varying workload, Kandasamy, *et al.* [KAN05] suggest a model predictive control approach to Dynamic Voltage Scaling. Lu, *et al.* [LU02b, LU03] use feedback control theory to design DVS algorithms for portable multimedia systems. The goal is to reduce decoding power consumption while maintaining a desired playback rate.

Soria-Lopez, *et al.* [SOR05] present an energy-based feedback control scheduling framework for power-aware soft real-time tasks executing in dynamic environments where real-time parameters are not previously know *a priori*. The framework contains an energy-based feedback scheduler and a power-aware optimization algorithm. Making use of a proportional controller based on an energy savings ratio, the feedback scheduler attempts to keep the CPU utilization at a high level to achieve high energy savings, and to distribute the computing resources among real-time tasks to maximize the control performance. The scheduler uses the energy feedback to calculate the amount of workload to be adjusted and provides the input for a Variable Voltage optimization Algorithm (VVA). The VVA algorithm is a greedy algorithm that adjusts the workload to minimize power consumption by computing a near optimal solution for the voltage scaling problem.

In a hard real-time environment, Zhu and Mueller [ZHU05a, ZHU05b] present a feedback DVS scheme using feedback control. The scheme can effectively schedule the energy for both static and dynamic workload. In order to make the DVS algorithm adaptable to changes in workload, Zhu [ZHU05b] integrates different feedback control structures. In addition, Zhu [ZHU05b] considers not only dynamic but also static power consumption caused by leakage current in circuits. The feedback DVS scheme is further extended accordingly.

1.3.2 Embedded Control Systems

The embedded control systems considered in the context of control and scheduling codesign are generally composed of a set of digital control loops. The control algorithm within each control loop is implemented separately as a real-time task. Naturally, the shared resource of interest is the CPU time. Accordingly, the most common objective of the codesign is to optimize the overall quality of control of the system subjected to CPU resource constraints [ARZ05a].

Seto, *et al.* [SET96, SET98] first consider optimizing the overall control performance under the constraint of real-time control task schedulability. In a preemptive, priority-based scheduling environment, a system performance index as a function of the sampling period is derived:

$$\Delta J(f) = \alpha e^{-\beta f}, \tag{1.1}$$

where α and β are related parameters, and f is the sampling frequency, the reciprocal of the sampling period.

Thereafter the above performance index is used in a lot of works to address the problem of overall control performance optimization under the system schedulability constraint. A variety of solutions have been proposed. For instance, Sha, *et al.* [CAC00, SHA00] consider the dynamic variations of task execution times with a similar system setup. The elastic feedback scheduling method is employed to attack the problem of task overrun, which delivers better performance than the WCET-based method. Taking into account the dynamic property of actual system responses, Lin, *et al.* [LIN03a] propose the notion of task importance to indicate the task's requirement on the sampling frequency. The CPU time is allocated in accordance with the online frequency-dependent performance change rate, thus achieving improvement over the optimization scheme in [SET96].

In order to optimize temporal attributes such as sampling period and delay under performance and resource constraints, Ryu, *et al.* [RYU97] present a heuristic approach to design real-time controllers by exploiting the period calibration method (PCM). The effects of both sampling period and delay on control performance are taken into account in [KIM98], where the author derives a performance index given by Eq.(1.2) that combines both of these two factors. The problem of task scheduling with feedback latency is described informally. A heuristic solution to this problem is also given.

$$J(h, \tau) = ce^{(ah + b\tau)} + d, \tag{1.2}$$

where h is the sampling period, τ is delay, a, b, c, and d are related parameters.

In [ABD97] Abdelzaher, *et al.* propose a model for QoS negotiation in building real-time services to meet both predictability and graceful degradation requirements. The efficiency and power of the model have been demonstrated for an automated flight control system. Following the fundamental idea of graceful gradation, Beccari, *et al.* [BEC99] describe a few elementary algorithms for sampling period adaptation.

The major topic of [EKE00] is also how to allocate the computing resource among a set of (independent) control loops such that the overall control performance is optimized. For each control loop, a Linear Quadratic (LQ) cost function is taken as the performance index, which is formulated as a function of the sampling period. An ideal optimization routine is given to exploit feedback scheduling. Cervin, *et al.* [CER00, CER02] present a feedback-feedforward scheduling structure. Based on the formulation in [EKE00], a linear and a quadratic approximate control cost functions are derived in [CER02], as given by Eqs.(1.3) and (1.4) respectively. Therein the authors also propose a near-optimal, simple rescaling method to reassign sampling periods.

$$J(h) = \alpha + \gamma h, \tag{1.3}$$

$$\mathrm{J(h)} = \alpha + \beta \mathrm{h}^2. \tag{1.4}$$

In [HEN05] the authors present a feedback scheduling strategy for a set of LQ-controller tasks that makes use of feedback from the process states. Given the current states of the controlled processes, the sampling periods are determined analytically to optimize the aggregate cost over the scheduler period. The computing resource is optimally distributed through relating the cost to both the sampling periods and the process states. With the same topic, Castane, *et al.* [CAS06] propose to determine online the optimal task periods considering the response over a finite time horizon of the processes controlled by arbitrary linear control laws. The solution of the optimization problem for the general case is computationally expensive, and hence an approximate heuristic procedure that is more suitable for online use has been developed. In [CER06] Cervin and Alriksson propose an online optimal scheduling scheme for non-preemptive control tasks. The objective of the scheme is to minimize a quadratic performance criterion for the overall system. The optimal scheduling decision at each point in time is a function of the states of the controlled processes. To make the scheme solve the real-world scheduling problems, the authors propose to use the relaxed dynamic programming technique to computer suboptimal solutions with error bounds.

Marti, *et al.* [MAR02a, MAR02b] illustrate in detail the concept of *flexible timing constraints*. The jitter compensation method that dynamically updates controller parameters is exploited. With the formulation of the QoC scheduling problem, the basic framework for integrating scheduling and control design and analysis is presented. In [MAR04] Marti, *et al.* point out that allocating CPU resource in accordance with the current states of controlled systems could effectively improve the overall control performance. Following this idea, an optimal resource allocation policy is developed. For multi-version control tasks, Lin, *et al.* [LIN04b] design a dynamic QoS manager in the face of discrete QoS levels. Using the method, the tasks are dynamically scheduled in an optimal way that the global benefit density is maximized.

Although tasks are often assumed to be independent in most work, the dependency constraints between tasks are studied in [YUA05], where a cascaded feedback scheduling method is presented. The method is mainly composed of a feedback controller and a frequency actuator. The former, based on the PID algorithm, takes the responsibility of controlling a specific performance index by regulating the CPU utilization of tasks. The latter then derives the sampling frequencies under the current utilization constraint using the task-oriented scheduling algorithm.

Codesign approaches based on advanced control or optimization techniques have also been presented. In [PAL02b] Palopoli, *et al.* formulate the problem of optimizing the robustness of the controlled systems under CPU resource constraints, assuming a time-triggered model of computation for control tasks and a real-time preemptive scheduling policy. Taking the sampling periods and the feedback gains as adjustable variables, an efficient numerical solution is obtained using a branch-and-bound algorithm. In [SIM05] the H_∞ control theory is used to design a robust feedback scheduler that is applied to a robot control system demonstrating good performance. A sampling period optimization method based on evolution strategy is suggested in [JIN05a]. A fuzzy feedback scheduling system is developed for real-time control tasks in [LEI05], where a fuzzy controller is used to determine the tasks' priorities dynamically. A feedback scheduler is used to assign the CPU utilization of tasks based on their priorities.

In most cases, the feedback scheduler regulates the system workload by means of sampling period adjustment. However, the changes in sampling period affect the performance of the control loop since control algorithms are often designed assuming constant sampling periods. To address this problem, Robert, *et al.* [ROB05] propose a sampling period dependent RST control algorithm that facilitates the applications of feedback scheduling.

In [ARZ99a] Årzén presents a simple, event-based PID controller to reduce the CPU resource consumption associated with the control task. The basic idea is to execute the PID control program only when necessary. It is argued that this makes possible large overhead reductions, while delivering control performance with only minor degradation in most situations.

In the above work the sampling periods of control tasks are selected as the scheduling parameters to be adjusted online. In contrast, adjusting the execution time rather than the sampling period of the control task may be a better approach to regulating the task utilization in cases where anytime control algorithms are used. Henriksson, *et al.* [HEN02] first study the feedback scheduling methods based on execution time adjustment. The MPC algorithm is treated as a kind of anytime control algorithms. The feedback information from multiple MPC optimization algorithms is exploited to decide when to stop the corresponding iterative optimization routines and to produce the control commands, as well as which task to execute at specific points in time.

Zhou, *et al.* [ZHO05b] propose to combine feedback scheduling and the constant bandwidth server (CBS) algorithm to schedule a set of MPC tasks. Each MPC task is allocated with a CBS, whose reserved processor time is adjusted online to deal with the time-varying execution times. Xia and Sun [XIA05b, XIA05c] study the scheduling of iterative optimal control tasks under dynamic resource constraints. For the case where the task execution time can be measured accurately, a simple iteration stopping scheme that is easy to implement is presented. The objective is to maximize the number of iterations while satisfying the system schedulability constraint. To address the inherent uncertainty in the execution of iterative control tasks, a fuzzy feedback scheduler is designed to re-assign online the maximum allowable number of iterations for the individual iterative control algorithms.

Still another possible approach to realize feedback scheduling is using tasks' deadlines as the manipulated variable. To minimize jitter, for instance, a feedback scheduling method featuring deadline adjustment based on PID control is presented in [ZHO05a].

There is also considerable interest in developing scheduling schemes better suited for Real-Time Control Systems. For instance, the integration of static cyclic scheduling and LQ optimal control is studied in [REH00], where an approach to obtaining the offline schedule that yields the minimum overall control cost is presented. Palopoli, *et al.* [PAL02c] develop a simulation tool RTSIM for Real-Time Control Systems design, as well as a reservation-based real-time task scheduling algorithm using constant bandwidth servers. To reduce computational delay and jitter, the methodology of scheduling subtasks rather than the original control tasks that could be divided into several subtasks is explored in e.g. [ALB00, CER99, CRE99, TIA06].

Based on the EDF algorithm along with a sporadic server, Liu and Fei [LIU05a] present an approach to calculating the optimal sampling frequencies. An optimal scheduling algorithm is also given to exploit dual priorities and a subtask model. Jin, *et al.* [JIN05b] present a threshold-based largest dedication first scheduling algorithm for control tasks characterized by fuzzy deadlines. Xia, *et al.* [XIA06a] propose the Maximum Urgency First (MUF) scheduling algorithm, on which a control-oriented direct feedback scheduling scheme is based.

Energy management plays an increasingly important role in the majority of embedded real-time systems. As a consequence, it becomes ever more crucial to save energy in embedded controllers, but very little work has been done in this direction.

Using a performance index that involves both control performance and energy consumption, Lee and Kim [LEE04] formulate the energy management problem in multitasking Real-Time Control Systems as an optimization problem. Both static and dynamic solutions are presented for the problem. The static solution obtains the optimal processor speed and a set of periods for given control tasks. The dynamic solution utilizes system services of real-time operating systems to overcome unavoidable deficiencies of the static solution and to further reduce the energy consumption of the overall system by switching the sampling periods between two values, based on whether or not the corresponding systems under control are in steady states.

In a way similar to [LEE04], Wang, *et al.* [WAN06] describe the problem of integrated optimization of sampling periods and CPU speed using a performance index related to both control cost and energy consumption. For the problem, a static solution based on evolution strategy is proposed.

Zhao and Xia [ZHA06a] present a highly cost-effective DVS method based on asynchronous period adjustment. Using the method, the sampling period of each task will be updated according to the current control performance of the relevant loop each time the task executes. The objective of the update is to reduce the amount of energy consumed by the task. In [ZHA06b] Zhao and Xia present a simple yet efficient DVS scheme that combines time-triggered and event-triggered mechanisms to attack the two typical classes of workload variations in Real-Time Control Systems.

1.4 Integrated Control and Communication

In the area of integrated control and computing, the bottleneck resource of the systems considered is the computing resource, while in the systems considered for integrated control and communication, the bottleneck resource demanding special attention will be the communication resource, e.g. network bandwidth. Just as done in the previous section, two categories of related work in the area of integrated control and communication are identified: control of networks and control over networks.

To avoid unexpected confusion, this book uses the term *network control* to represent control of networks, and *networked control* to represent control over networks. Furthermore, to distinguish between wired networks and wireless networks, the control systems closed over networks are classified into (wired) Networked Control Systems (NCSs) and Wireless Control Systems (WCSs).

1.4.1 Control of Networks

The area of network control is quite broad, involving a variety of aspects of network communications [ARZ06a, MUR03]. However, this book is only concerned with the application of the feedback scheduling methodology to communication resource management.

The most representative application of control theory and technologies to wired networks is network congestion control [XU06a], for example, the end-to-end rate control of IP networks and the Active Queue Management (AQM) in routers. In history, the design of traditional congestion control algorithms in networks, often done by system engineers, is rarely related to feedback control theory. Since Kelly, *et al.* [KEL98] made important progress in the theoretical foundation of congestion control, a significantly large number of results in network congestion control based on feedback control theory have been reported recently [ARZ05b, ARZ06a]. To maintain consistency, only some of those efforts that are closely related to this work will be summarized below.

The classical control techniques are successfully applied to active queue management by Hollot, *et al.* in [HOL01], where a proportional controller and a PI controller are designed and analysed. Deng, *et al.* [DEN03] add the derivative component onto the PI controller, thus forming the more general PID framework. To improve the stability and robustness of the system, the authors present an adaptive control approach to AQM. Ren, *et al.* [REN02] design an active queue manager using fuzzy logic control, which delivers better performance than the PI control based approach.

Compared with wired networks, the resource management in wireless environments could be much more difficult. This is mainly because wireless networks transmit data over radio channels, which is characterized by considerably higher level of non-determinism [ARZ05b, ARZ06b]. The channel capacities of wireless networks depend heavily on many factors such as transmit power, bandwidth and the size of time slots [ARZ06a], and thereby are potentially variable, which is in contrast to wired networks that always hold fixed link capacities.

Besides the bandwidth, the resources in wireless networks that need to be managed also include the transmit power. Energy consumption has now become a critical factor that directly affects the lifetime of many systems, especially for e.g. large scale wireless sensor networks [DAI05, DAI06, LIU05b]. Intuitively, the transmit power influences not only the channel capacities of wireless networks, but also the energy consumption of the nodes. Therefore, from a resource management point of view, the transmit power control in wireless networks is certainly an important issue.

Adaptive resource management in mobile wireless networks is studied in [HOS04a], where the authors design a proportional controller and a PI controller to maintain the handoff failure rate below a desired level. In [HOS04b] a generalized feedback-based control algorithm is presented for the problem, together with two improvement methods. Chen and Chen [CHE06] propose a closed loop scheduling scheme for wireless mobile communication services that require dynamic QoS, by combining an open-loop scheduler, a QoS probe and a PID controller.

Priscoli and Isidori [PRI05] design a traffic control module for wireless IP networks using control engineering methods. The congestion control and scheduling problem is dealt with in a highly integrated and feedback-based manner, while taking into account current network workload. In wireless ad hoc environments, He, *et al.* [HE05] present a feedback control based resource allocation scheme, where the wireless link queue length is controlled by dynamically manipulating the sending rate of the resource.

Abdelzaher, *et al.* [ABD04] attempt to apply feedback control to the area of wireless sensor networks. For this purpose, an analytic model is constructed to describe the relationships between timeliness, energy, and the degree of data aggregation, as well as to quantify constraints stemming from real-time requirements. A feedback controller is designed to adapt the degree of data aggregation online in response to network workload conditions, whose objective is to maintain specified acceptable latency bounds on data delivery while minimizing the energy consumption.

1.4.2 Networked Control Systems

Networked control systems closed over wired networks are a natural result of the integration of control and communication. In modern networked control systems control applications depend either on traditional control networks (i.e. fieldbuses [THO05, TOV99]) or on general-purpose data networks such as Ethernet and Internet to transmit data and exchange information. Significant progress has been made in this area since Halevi and Ray [HAL88, RAY88] introduced the concept of integrated communication and control systems in late 1980s.

Roughly speaking, the majority of existing efforts on networked control systems may be classified into three categories:

(1) controller design based approaches,

(2) network design based approaches, and

(3) control and network codesign.

The key to controller design based approaches is to design control algorithms that are robust against network-induced delays, packet losses and jitters. Examples of this category include [LI05, NIL98, TAT04, ZHA01]. The fundamental technical routine is as follows: first collect the distributions of delay and packet loss through theoretical and/or experimental analysis of the temporal behaviour of networks with specified system structure and communication protocols, or just simply assume that the distributions of delay and packet loss are known *a priori*; and then, with respect to the specific properties of the delay and packet loss, use appropriate control techniques such as stochastic optimal control and hybrid control to design control algorithms robust to these delays, packet losses and jitters.

The approaches based on network design mainly deal with how to improve network QoS such that the timing requirements of the control systems are met, e.g. [LIA01]. The general idea behind is to reduce network-induced delay, packet loss and jitter as much as possible, by properly designing network structures, communication protocols, bandwidth allocation algorithms, media access control schemes, etc. When doing this, one should keep in mind those factors associated with the networks that determine the delay, packet loss and jitter.

Generally, substantial interaction between control theory and communication technology is rarely involved in the first two categories. This book is concerned with the third category, i.e. the codesign of control and network. Below, some work falling into this category is outlined with emphasis on network resource management.

In [HRI02] Hristu-Varsakelis and Kumar propose a feedback communication policy for a group of linear control systems closed over an idealized shared network. The policy is used to decide which system should be admitted into the network and for how long. The use of feedback in making communication decisions results in a set of autonomous dynamical systems which are coupled to one another due to the presence of communication constraints. For CAN (Controller Area

Network) based control systems, Walsh and Ye [WAL01] propose the maximum-error-first with try-once-discard (MEF-TOD) scheduling algorithm. Yepez, *et al.* [YEP02, YEP03] present the Large Error First (LEF) scheduling algorithm for control loops. The basic idea of the algorithm is to schedule first the packets belonging to the control loop whose current absolute control error is the maximum.

Ren, *et al.* [REN03] study the QoS management of networked control systems that takes advantage of parallel processing, and present the most-vacancy-time-first algorithm for message scheduling. Xia, *et al.* [XIA05a, ZHA06c] design QoS managers using neural network based predictors. Utilizing the prediction of network workload, the QoS manager adapts the sampling periods of control loops, thus attacking the impact of variable workload on control performance.

In [HON00] Hong and Kim present a bandwidth allocation scheme for the CAN bus. The scheme not only satisfies the performance requirement of real-time application systems over the CAN bus, but also fully utilizes the bandwidth. Kim and Park [KIM02] develop a heuristic method for determining sampling periods and priorities, while taking into account the end-to-end timing constraints in networked control systems. Park, *et al.* [KIM03, PAR02] present another heuristic method for control network scheduling. The method can adjust the sampling period as small as possible, allocate the bandwidth of the network for three types of data (i.e. periodic data, sporadic data, and messages), and exchange the transmission orders of data for sensors and actuators. It can guarantee real-time transmission of sporadic and periodic data, and minimum utilization for non-real-time messages.

Zhang [ZHA01] and Branicky, *et al.* [BRA02] introduce the methodology of integrated control and scheduling into the network scheduling of multi-loop network control systems. Some basic methods are also given. Assuming ideal networks, the optimization problem of network scheduling is formulated. The basic approach to solving the optimal sampling periods is exemplified. The system stability conditions in the presence of packet losses are derived. The idea of coping with overload through intentionally discarding some packets is also illustrated. Following the approach, He, *et al.* [HE04] propose to obtain the optimal sampling periods using the genetic algorithm technology. Another solution based on nonlinear optimization is given in [ALH05].

Velasco, *et al.* [VEL04] present a dynamic approach to bandwidth management in networked control systems that allows control loops to consume bandwidth according to the dynamics of the controlled process while attempting to optimize overall control performance. The original state-space representation of each controlled process is augmented with a new state variable that describes the network dynamics. With the approach, the allocation of bandwidth to control loops can be done locally at runtime according to the state of each controlled process without causing overload situations, and control laws can be designed to account for the variations in the assigned bandwidth preventing unexpected control performance degradation and even destabilization that would otherwise occur.

Gravagne, *et al.* [GRA04b] propose to dynamically adjust the sampling periods under the system stability constraint, so as to reduce the applications' bandwidth requirements. Bai, *et al.* [BAI05] present a jitter-dependent optimal bandwidth scheduling algorithm to make tradeoffs between bandwidth usage and system performance. Tipsuwan and Chow [TIP04] present an approach to enabling existing controllers for networked control and tele-operation, which uses middleware to modify the output of an existing controller based on the gain scheduling technique with respect to the current network traffic conditions.

1.4.3 Wireless Control Systems

Recent years have witnessed the emergence of wireless networking technologies that are ever-increasingly mature, the availability of a large number of commercialized supporting devices that are more inexpensive, as well as widespread deployment of these technologies in a variety of systems that are influencing (or will influence) many aspects of our every-day lives. All of these make it possible to realize real-time control over wireless networks. Indeed, the field of wireless control is continuously attracting attention from both academia and industry [AND05, ARZ06b, COL05, MAT05, RAM05, SAN05, WIL05], for its incomparable advantages over traditional wired networked control systems, including e.g. full support for node mobility, easier installation and maintenance, and cheaper costs. However, it is highly challenging to accomplish effective control based on wireless networks (e.g. WLAN, ZigBee, and Bluetooth) that are much less dependable than wired networks. At this moment this area is just in its infancy.

Ye, *et al.* [YE01] propose a prioritized CSMA/CA (Carrier Sense Multiple Access with Collision Avoidance) protocol for real-time wireless local area networking. The protocol, based on the IEEE 802.11 wireless standard, mixes real-time traffic with standard multimedia data in a way that the loop stability is assured. Under the framework several algorithms for dynamically scheduling the traffic of wireless control systems, i.e. the constant penalty, estimated error order and lag first order schemes are proposed and validated.

The impact of varying fading wireless channels on control performance is studied in [MOS04], where the authors suggest that the controller parameters should be dynamically adapted with respect to channel conditions. Xiao, *et al.* [XIA03] develop an offline approach to optimize the stationary performance of a linear control system by jointly allocating communication resources and tuning parameters of the controller.

Liu and Goldsmith [LIU04] introduce the cross-layer design methodology into the joint design of wireless network and distributed controllers. With the goal of optimizing the control performance, a four-layer framework is presented. A numerical example is given to show how the link layer, the MAC (Medium Access Control) layer and the selection of sampling periods affect the control performance. In addition, an iterative procedure for cross-layer design of wireless control systems is illustrated.

Ploplys, *et al.* [PLO04] examine the performance of WLAN when the UDP protocol rather than the usual TCP protocol is employed for communication among nodes. The authors propose a method to modify the sampling period of the control system to adapt to varying network conditions, in which a PI control algorithm is used. Using the same control structure, Kawka and Alleyne present another heuristic algorithm to adapt the sampling period in [KAW05].

Tzes, *et al.* [TZE03] present a client-centric mobile wireless control architecture, where the feedback control loop is closed over a General Purpose Radio Service (GPRS) communication channel. The characteristics and effects of varying delays induced by mobile wireless communications are studied experimentally. A corresponding controller is designed using the LMI theory that provides the worst-case scenario of the delay that the controller can tolerate. In [PAN04] the authors present an integrated optimization framework for realtime control applications over WLAN. The framework monitors the network QoS, adjusts the maximum number of allow-able data retransmission, attempts to optimize wireless transmission, and periodi-cally tunes the controller's parameters to improve control performance. Nikolak-opoulos, *et al.* [NIK05] present a gain scheduling method for LQR (Linear Quadratic Regulator) controllers to compensate for delay induced by multihop communication in wireless sensor networks running on top of the IEEE 802.11b protocol.

1.5 Perspective on Feedback Scheduling

Table 1.2 roughly summarizes the majority of aforementioned feedback scheduling methods in the field of control and scheduling codesign. Classified with respect to the main technique used as well as the application area, only those methods closely related to this work are included here.

Table 1.2 Different types of existing feedback scheduling schemes

Type	Application area				
	ECS	DVS	NCS	WCS	Others
Heuristic	[BEC99][CER00] [CER02][LIN04b]		[GRA04b][PAR02] [REN03][VEL04]	[KAW05][MOS04] [PAN04]	
PID	[YUA05][ZHO05a]	[LU03] [SOR05] [VAR03] [ZHU05b]		[PLO04]	[ABE02][CHE06][GOE04] [HOL01][HOS04a][LIN04a] [LU02a][TON04][ZHA04]
Optimal	[CAS06][CER06] [EKE00][HEN05] [JIN05a][LIN03a] [MAR04][SET96]	[LEE04] [WAN06]	[ALH05][BRA02] [HE04]		[LAW01]
Adaptive					[ABE00][DEN03][LIU06] [WEI03][WEI05]
Predictive		[KAN05]			[LU05][WAN05][XU06b]
Fuzzy	[LEI05]				[DIA02][JIN04][RAV01] [REN02]
Others	[SIM05]	[ZHA06a]	[HRI02][WAL01] [YEP02]		[DU04][KOU05][PAL03] [SAH02]

There is an obvious emergence of the field of feedback scheduling following the trend towards convergence of computing, communication, and control. The methodology of feedback scheduling has been recognized as a promising approach to flexible resource management and used by more and more researchers from diverse disciplines. An exciting fact is that the number of applications of feedback scheduling theory and techniques to control systems especially embedded control systems is rapidly growing. The trend is to constitute a fresh and multidisciplinary branch of academic research in real-time systems [SHA04].

Regardless, feedback scheduling is a newly-emerging area. Currently the results in this area are very limited, as compared with the established rich knowledge repositories associated with real-time control, real-time computing, and real-time communication. Furthermore, a lot of work in the area of control and scheduling codesign deals with general-purpose computing/communication systems rather than control applications. A consequence is that so far very little work has been done with feedback scheduling strategies dedicated to Real-Time Control Systems. This is particularly the case for areas such as Dynamic Voltage Scaling, networked control systems, and wireless control systems.

On the one hand, only a small number of control principles and methods have been exploited for the purpose of feedback scheduling. Furthermore, many advantages of them have not been explored in these applications. On the other hand, quite a large number of resource management problems that exist in real-world real-time systems (particularly resource constrained Real-Time Control Systems) are still open. These problems include for example the measurement noise in system parameters, the unpredictability of task execution times, the unavailability of some necessary information, the variability of CPU/network workload, etc. In addition, there is a lack of a unified framework that can serve as a technological platform for the development of new theory and methods in this field.

To address the many problems associated with the implementation of Real-Time Control Systems in the new technological environment, more systematic investigations into codesign of control and scheduling are demanded. To meet this requirement, this work is endowed with the long-term goal of establishing the theoretical and technological framework of feedback scheduling for Real-Time Control Systems. In this book, a set of flexible resource management schemes will be developed within the framework of feedback scheduling to maximize the overall system performance of Real-Time Control Systems in the presence of uncertainties.

Beyond this work, there are many foreseeable challenges and opportunities facing the area of control and scheduling codesign, of which most are related to the design, analysis, and/or implementation of feedback schedulers. Given below are a few examples that provide some insights into these challenges and opportunities.

(1) *Control theoretic design and analysis of feedback schedulers.* As a mature discipline, feedback control provides engineering systems with a powerful theoretical foundation. Using feedback control theory to design and analyse feedback schedulers yields predictable feedback scheduling performance that

is easy to evaluate. The most challenging issue here is the modelling of the computing/communication resource scheduling system, which is a prerequisite for applying feedback control.

(2) *Feedback scheduling based on intelligent computing.* The target application environments of feedback scheduling are usually characterized by the presence of uncertainty. In many situations the accurate or approximate (first-principles) models for the resource scheduling systems are difficult to derive. In this context designing the feedback schedulers based on intelligent computing theory and techniques (e.g. fuzzy control) is an effective and promising approach. For instance, a possible topic is applying fuzzy feedback scheduling to the communication resource management in wireless control systems.

(3) *Feedback scheduling of complex control systems.* The complexity of many real-world control systems is quite high. For instance, some information may need to be exchanged between different control loops; there may be multiple sub-tasks within a control loop that must be implemented separately and be coordinated; more than one sampling node (i.e. sensor) may be used. These properties make it more difficult to design and implement feedback schedulers. If the control tasks have precedence constraints, from a real-time scheduling point of view, the scheduling of these tasks could become considerably complex. Also, it could become very hard to assign appropriate timing attributes for individual control tasks.

(4) *Controller design and analysis in the context of feedback scheduling.* The application of feedback scheduling to control systems gives rise to dynamic variations in some design parameters (e.g. sampling periods) of the control algorithms, which may possibly deteriorate control performance to some degree. Therefore, it is sometimes necessary to modify the original system design, for example, to update online the controller parameters with respect to the changes in sampling periods and/or delay. However, this complicates theoretical analysis of the resulting control performance. Of course, it is also possible to directly design such control algorithms that are robust enough against these changes.

(5) *The practical implementation of feedback schedulers.* To realize the potential of feedback scheduling, methods and tools for implementing feedback schedulers in real systems must be developed. Some technical problems associated with practical development platforms should be addressed. For example, a non-trivial question is how to exchange information between the feedback scheduler and relevant entities within control loops, especially in networked environments.

References

[ABD97] T.F. Abdelzaher, E.M. Atkins, K.G. Shin. QoS Negotiation in Real-Time Systems and Its Application to Automated Flight Control. *Proc. of the Third IEEE Real-Time Technology and Applications Symposium (RTAS)*, Montreal, Canada, pp.228 – 238, 1997.

[ABD04] T.F. Abdelzaher, T. He, J. Stankovic. Feedback Control of Data Aggregation in Sensor Networks. *43rd IEEE Conf. on Decision and Control (CDC)*, Paradise Island, Bahamas, pp.1490 – 1495, 2004.

[ABE00] L. Abeni, L. Palopoli, G. Buttazzo. On Adaptive Control Techniques in Real-Time Resource Allocation. *Proc. ECRTS*, Sweden, pp.129 – 136, 2000.

[ABE02] L. Abeni, L. Palopoli, G. Lipari, J. Walpole. Analysis of a Reservation-Based Feedback Scheduler. *Proc. 23rd IEEE RTSS*, Austin, Texas, pp.71 – 80, 2002.

[ABE05] L. Abeni, T. Cucinotta, G. Lipari, L. Marzario, L. Palopoli. QoS Management Through Adaptive Reservations. *Real-Time Systems*, Vol. 29, No. 2 – 3, pp.131 – 155, 2005.

[ALB00] P. Albertos, A. Crespo, I. Ripoll, M. Vallés, P. Balbastre. RT control scheduling to reduce control performance degrading. *IEEE Conf. on Decision and Control*, Sydney, Vol.5, pp.4889 – 4894, 2000.

[ALH05] A. Al-Hammouri, V. Liberatore. Optimization Congestion Control for Networked Control Systems. *IEEE INFOCOM'05 Student Workshop*, 2005.

[ALO03] R. Al-Omari, G. Manimaran, M.V. Salapaka, A.K. Somani. Novel Algorithms for Open-Loop and closed loop Scheduling of Real-Time Tasks in Multiprocessor Systems Based on Execution Time Estimation. *Int. Symposium on Parallel and Distributed Processing (IPDPS)*, 2003.

[ALT02] K. Altisen, G. Goessler, J. Sifakis. Scheduler Modeling Based on the Controller Synthesis Paradigm. *Real-Time Systems*, Vol.23, pp.55 – 84, 2002.

[AMI05] M. Amirijoo, J. Hansson, S. Gunnarsson, S. H. Son. Enhancing Feedback Control Scheduling Performance by On-line Quantification and Suppression of Measurement Disturbance. *Proc. 11th IEEE RTAS*, California, USA, pp.2 – 11, 2005.

[AND05] M. Andersson, D. Henriksson, A. Cervin, K.E. Årzén. Simulation of Wireless Networked Control Systems. *Proc. of 44th IEEE Conf. on Decision and Control and European Control Conf.*, Seville, Spain, pp.476 – 481, 2005.

[ANT99] P. Antsaklis, T. Basar, R. DeCarlo, N. H. McClamroch, M. Spong, S. Yurkovich. Report on the NSF/CSS Workshop on New Directions in Control Engineering Education. *IEEE Control Systems Magazine*, Vol.19, No.5, pp.53 – 58, 1999.

[ARZ99a] K.E. Årzén. A Simple Event Based PID Controller. *Proc. 14th IFAC World Congress*, Beijing, China, pp.423 – 428, 1999.

[ARZ99b] K.E. Årzén, B. Bernhardsson, J. Eker, A. Cervin, P. Persson, K. Nilsson, L. Sha. Integrated Control and Scheduling. *Research Report*, ISSN 0820 – 5316, Dept. Automatic Control, Lund Institute of Technology, 1999.

[ARZ03] K. E. Årzén, A. Cervin, D. Henriksson. Resource-Constrained Embedded Control Systems: Possibilities and Research Issues. *Proc. of Codesign of Embedded Real-Time Systems Workshop*, Porto, Portugal, 2003.

[ARZ05a] K.E. Årzén, A. Cervin. Control and Embedded Computing: Survey of Research Directions. *Proc. 16th IFAC World Congress*, Prague, Czech Republic, 2005.

[ARZ05b] K. E. Årzén (Editor). ARTIST2 Strategic Research Agenda on Control for Embedded Systems. *Lund Workshop on Control for Embedded Systems*, Lund, 2005.

[ARZ06a] K.E. Årzén, A. Robertsson, D. Henriksson, M. Johansson, H. Hjalmarsson, K.H. Johansson. Conclusions of the ARTIST2 Roadmap on Control of Computing Systems. *ACM SIGBED Review*, Vol.3, No.3, pp.11 – 20, 2006.

[ARZ06b] K.E. Årzén, A. Bicchi, S. Hailes, K. H. Johansson, J. Lygeros. On the Design and Control of Wireless Networked Embedded Systems. *Proc. of IEEE Computer Aided Control Systems Design Symposium*, pp.440 – 445, 2006.

[AYD04] H. Aydin, R. Melhem, D. Mosse, P. Mejia-Alvarez. Power-Aware Scheduling for Periodic Real-Time Tasks. *IEEE Trans. Computers*, Vol. 53, No.5, pp.584 – 600, 2004.

[BAI05] T. Bai, Z. Wu, G. Yang. Optimal Bandwidth Scheduling of Networked Control Systems (NCSs) in Accordance with Jitter. *Journal of Zhejiang University SCIENCE*, Vol. 6A, No.6, pp.535 – 542, 2005.

[BEC99] G. Beccari, S. Caselli, M. Reggiani, F. Zanichelli. Rate Modulation of Soft Real-Time Tasks in Autonomous Robot Control Systems. *11th Euromicro Conf. on Real-Time Systems*, York, UK, pp.21 – 28, 1999.

[BEN06] M.M.B. Gaid. Optimal Scheduling and Control for Distributed Real-time Systems.Ph.D Thesis, COSI Laboratory, ESIEE, Paris, 2006.

[BOU05] B. Bouyssounouse, J. Sifakis. *Embedded Systems Design: The ARTIST Roadmap for Research and Development.* Springer, 2005.

[BRA02] M.S. Branicky, S.M. Philips, W. Zhang. Scheduling and Feedback Codesign for Networked Control Systems. *Proc. 41st IEEE CDC*, Las Vegas, pp.1211 - 1217, 2002.

[BUT02] G. Buttazzo, L. Abeni. Adaptive Workload Management Through Elastic Scheduling. *Real-Time Systems*, Vol. 23, No.1/2, pp.7 - 24, 2002.

[BUT04] G. Buttazzo, M. Velasco, P. Martí, G. Fohler. Managing Quality-of-Control Performance Under Overload Conditions. *16th Euromicro Conf. on Real-Time Systems (ECRTS04)*, Catania, Italy, pp.53 - 60, 2004.

[CAC00] M. Caccamo, G. Buttazzo, L. Sha. Elastic feedback control. *12th Euromicro Conf. on Real-Time Systems*, Stockholm, Sweden, pp.121 - 128, 2000.

[CAS06] R. Castañé, P. Martí, M. Velasco, A. Cervin, D. Henriksson. Resource Management for Control Tasks Based on the Transient Dynamics of Closed Loop Systems. *18th Euromicro Conf. on Real-Time Systems*, Dresden, Germany, 2006.

[CER99] A. Cervin. Improved Scheduling of Control Tasks. *Proc. 11th ECRTS*, York, UK, pp.4 - 10, 1999.

[CER00] A. Cervin, J. Eker. Feedback Scheduling of Control Tasks. *Proc. 39th IEEE CDC*, pp.4871 - 4876, 2000.

[CER02] A. Cervin, J. Eker, B. Bernhardsson, K.E. Årzén. Feedback-Feedforward Scheduling of Control Tasks. *Real-Time Systems*, Vol. 23, No.1, pp.25 - 53, 2002.

[CER06] A. Cervin, P. Alriksson. Optimal On-line Scheduling of Multiple Control Tasks: A Case Study. *Proc. of the 18th Euromicro Conf. on Real-Time Systems*, Dresden, Germany, 2006.

[CHE06] J.L. Chen, N.K. Chen. Feedback Closed Loop Scheduling Discipline for QoS Guarantee in Mobile Applications. *Wireless Networks*, Vol. 12, No. 2. pp.223 - 232, 2006.

[COL05] J. Colandairaj, G.W. Irwin, W.G. Scanlon. Analysis of an IEEE 802.11b Wireless Networked Control System. *Proc. of the 16th IFAC World Congress*, Prague, Czech Republic, 2005.

[CRE99] A. Crespo, I. Ripoll, P. Albertos. Reducing Delays in RT Control: The Control Action Interval. *Proc. IFAC World Congress*, Beijing, 1999.

[CUC04] T. Cucinotta, L. Palopoli, L. Marzario, G. Lipari, L. Abeni. Adaptive Reservations in a Linux Environment. *Proc. 10th IEEE RTAS*, Toronto, Canada, pp.238 - 245, 2004.

[DAI05] X.H. Dai, F. Xia, Z. Wang, Y.X. Sun. A Survey of Intelligent Information Processing in Wireless Sensor Networks, *MSN'05*, *Lecture Notes in Computer Science*, Vol. 3794, pp.123 - 132, 2005.

[DAI06] X.H. Dai, F. Xia, Z. Wang, Y.X. Sun. An Energy-Efficient In-Network Aggregation Query Algorithm for Wireless Sensor Networks. *Proc. of the 1st IEEE Int. on Innovative Computing, Information and Control (ICICIC'06)*, Beijing, China, pp.255 – 258, 2006.

[DEN03] X. Deng, S. Yi, G. Kesidis, C. R. Das. A Control Theoretic Approach for Designing Adaptive AQM Schemes. *Proc. of IEEE Global Telecommunications Conf. (GLOBECOM'03)*, San Francisco, pp.2947 – 2951, 2003.

[DIA02] Y. Diao, J.L. Hellerstein, S. Parekh. Optimizing Quality of Service Using Fuzzy Control. *Proc. 13th IFIP/IEEE DSOM*, pp.42 – 53, Springer-Verlag, 2002.

[DOD06] U.S. Department of Defense (DoD) News. Universities Selected for Research Funding. http://www.defenselink.mil/releases/2006/nr20060222 – 12548.html, 2006.

[DU04] B. Du, D.C. Levy. On the Robust Capability of Feedback Scheduling in ORB Middleware. *5th Australian Workshop on System and Software Architectures (AWSA'04)*, Melbourne, 2004.

[EKE00] J. Eker, P. Hagander, K.E. Årzén. A Feedback Scheduler for Real-Time Controller Tasks. *Control Engineering Practice*, Vol. 8, No.12, pp.1369 – 1378, 2000.

[GOE04] A. Goel, J. Walpole, M.H. Shor. Real-Rate Scheduling. *Proc. of the IEEE Real-Time Technology and Applications Symposium (RTAS)*, pp.434 – 441, 2004.

[GRA03] S. Graham, P.R. Kumar. The Convergence of Control, Communication, and Computation. *Lecture Notes in Computer Science*, Springer-Verlag, Vol. 2775, pp.458 – 475, 2003.

[GRA04a] S. Graham, G. Baliga, P.R. Kumar. Issues in the Convergence of Control With Communication and Computing: Proliferation, Architecture, Design, Services, and Middleware. *Proc. of the 43rd IEEE Conf. on Decision and Control*, Bahamas, Vol. 2, pp.1466 – 1471, 2004.

[GRA04b] I.A. Gravagne, J.M. Davis, J.J. Dacunha, R.J. Marks. Bandwidth Reduction for Controller Area Networks Using Adaptive Sampling. *IEEE Int. Conf. on Robotics and Automation*, Vol. 5, pp.5250 – 5255, 2004.

[HAL88] Y. Halevi, A. Ray. Integrated communication and Control Systems: Part I – Analysis. *ASME Journal of Dynamic Systems, Measurement and Control*, Vol. 110, pp.367 – 373, 1988.

[HE04] J. He, H. Zhang, Y. Jing. An Integrated Control and Scheduling Optimization Method of Networked Control Systems. *Journal of Electronic Science and Technology of China*, No.02, pp.57 – 60, 2004.

[HE05] W. He, X. Liu, K. Nahrstedt. A Feedback Control Scheme for Resource Allocation in Wireless Multi-hop Ad Hoc Networks. *Proc. of ACM MobiQuitous Conf.* , 2005.

[HEN02] D. Henriksson, A. Cervin, J. Åkesson, K.E. Årzén. Feedback Scheduling of Model Predictive Controllers. *Proc. 8th IEEE RTAS*, San Jose, CA, pp.207 – 216, 2002.

[HEN05] D. Henriksson, A. Cervin. Optimal On-line Sampling Period Assignment for Real-Time Control Tasks Based on Plant State Information. *Proc. of the 44th IEEE Conf. on Decision and Control and European Control Conf. ECC*, Seville, Spain, 2005.

[HEN06] T. Henningsson, A. Cervin. Event-based Control over Networks: Some Research Questions and Preliminary Results. *Proc. of Reglermöte'06*, Stockholm, Sweden, 2006.

[HEL04] J. L. Hellerstein, Y. Diao, S. Parekh, D. Tilbury. Feedback Control of Computing Systems. *Wiley-Interscience*, 2004.

[HEL05] J. L. Hellerstein, Y. Diao, S. Parekh, D. Tilbury. Control Engineering for Computing Systems. *IEEE Control Systems Magazine*, Vol.25, No.6, pp.56 – 68, 2005.

[HOL01] C.V. Hollot, V. Misra, D. Towsley, W. Gong. On Designing Improved Controllers for AQM Routers Supporting TCP Flows. *Proc. of INFOCOM*, Anchorage, Alaska, pp.1726 – 1734, 2001.

[HON00] S. H. Hong, W.H. Kim. Bandwidth Allocation Scheme in CAN Protocol. *IEE Proceedings—Control Theory and Applications*, Vol. 147, No.1, pp.37 – 44, 2000.

[HOS04a] M. Hossain, M. Hassan, H. Sirisena. Adaptive Resource Management in Multi-Service Mobile Wireless Cellular Networks Using Feedback Control. *Proc. of IEEE Vehicular Technology Conf.* , New Jersey, USA, pp.194 – 199, 2004.

[HOS04b] M. Hossain, M. Hassan, H. Sirisena. Adaptive Resource Management in Mobile Wireless Networks Using Feedback Control Theory. *Telecommunications Systems*, pp.40 – 415, 2004.

[HRI02] D. Hristu-Varsakelis, P.R. Kumar. Interrupt-based Feedback Control over a Shared Communication Medium. *IEEE Conf. on Decision and Control*, pp.3223 – 3228, 2002.

[JIN04] H. Jin, H.A. Wang, Y. Fu, Q. Wang, H. Wang. A Fuzzy Feedback Control Real-Time Scheduling Algorithm. *Chinese Journal of Software*, Vol.15, No.6, pp.791 – 798, 2004.

[JIN05a] H. Jin, H. Wang, H. A. Wang, G. Dai. Optimization Design of Controller Periods Using Evolution Strategy. *CIS'05*, *Lecture Notes in Artificial Intelligence*, Vol. 3801, pp.1100 – 1105, 2005.

[JIN05b] H. Jin, H.A. Wang, H. Wang, D. Wang. Scheduling Control Tasks with Threshold-based Largest Dedication First. *International Journal of Information Technology*, Vol.11, No.5, pp.61 – 70, 2005.

[KAN05] N. Kandasamy, S. Abdelwahed, G. Sharp, J. Hayes. An Online Control Framework for Designing Self-Optimizing Computing Systems: Application to Power Management. *Lecture Notes in Computer Science*, Vol.3460, pp.174 – 188, 2005.

[KAW05] P.A. Kawka, A.G. Alleyne. Stability and Feedback Control of Wireless Network Systems. *Proc. of the American Control Conf. (ACC05)*, Portland, OR, 2005.

[KEL98] F. Kelly, A. Maulloo, D. Tan. Rate Control in Communication Networks: Shadow Prices, Proportional Fairness and Stability. *Journal of the Operational Research Society*, Vol.49, pp.237 – 252, 1998.

[KER05] M. Kersten. Dynamic Voltage Scaling and its Scheduling Implications. *Research Seminar on Energy – Awareness*, Dept. of Computer Sci., Univ. of Helsinki, http://www. cs. helsinki. fi/u/kraatika/Courses/sem05a-EnergyAwareness.html. 2005.

[KIM98] B. Kook Kim. Task Scheduling with Feedback Latency for Real-Time Control Systems. *5th Int. Conf. on Real-Time Computing Systems and Applications (RTCSA)*, Hiroshima, Japan, pp.37 – 41, 1998.

[KIM02] H.Y. Kim, H.S. Park. Optimal Period and Priority Assignment for Guaranteeing Real-Time Requirements in Distributed Control Systems. *The 4th Asian Control Conf. (ASCC'02)*, pp.1052 – 1057, 2002.

[KIM03] D.S. Kim, Y.S. Lee, W.H. Kwon, H.S. Park. Maximum Allowable Delay Bounds of Networked Control Systems. *Control Engineering Practice*, Vol.11, pp.1301 – 1313, 2003.

[KOU05] X. Koutsoukos, R. Tekumalla, B. Natarajan, C. Lu. Hybrid Supervisory Utilization Control of Real-Time Systems. *IEEE Real-Time and Embedded Technology and Applications Symposium (RTAS'05)*, 2005.

[LAW01] D. Lawrence, J. Guan, S. Mehta, L. Welch. Adaptive Scheduling via Feedback Control for Dynamic Real-Time Systems. *Proc. 20th IEEE Int. Performance, Computing, and Communications Conf.*, Arizona, 2001.

[LEE04] H.S. Lee, B.K. Kim. Design of Digital Control Systems with Dynamic Voltage Scaling. *10th IEEE Real-Time and Embedded Technology and Applications Symposium (RTAS'04)*, 2004.

[LEI05] F. Lei, Y. Zou, J. Zhou. A Fuzzy Feedback Scheduling System for Real-Time Control Tasks. *Chinese Journal of Information and Control*, Vol. 34, No.5, pp.585 – 590, 2005.

[LI05] S.B. Li. Study on Robust Control Algorithm of Networked Control Systems. Ph.D Thesis, Zhejiang University, 2005.

[LIA01] F.L. Lian. Analysis, Design, Modeling, and Control of Networked Control Systems. Ph.D Thesis, University of Michigan, 2001.

[LIN03a] Q. Lin, P.C.Y. Chen, P.A. Neow. Dynamical Scheduling of Digital Control Systems. *IEEE Int. Conf. on Systems, Man and Cybernetics*, Vol.5, pp.4098 – 4103, 2003.

[LIN03b] S. Lin, G. Manimaran. Double-loop Feedback-based Scheduling Approach for Distributed Real-Time Systems. *Lecture Notes in Computer Science*, Vol. 2913, pp.268 – 278, 2003.

[LIN04a] S. Lin, G. Manimaran, B. Steward. Feedback-based Real-Time Scheduling in Autonomous Vehicle Systems. *Proc. 10th IEEE RTAS*, Toronto, Canada, pp.316 – 323, 2004.

[LIN04b] C. Lin, P. Marti, S.A. Brandt, S. Banachowski, M. Velasco, J.M. Fuertes. Improving Control Performance Using Discrete Quality of Service Levels in a Real-Time System. *10th IEEE Real-Time and Embedded Technology and Applications Symposium*, *Work in Progress Proc.*, 2004.

[LIU73] C. Liu, J. Layland. Scheduling Algorithms for Multiprogramming in a Hard Real-Time Environment. *Journal of the ACM*, Vol.20, pp.46 – 61, 1973.

[LIU00] J.W.S. Liu. *Real-Time Systems*, Prentice Hall, 2000.

[LIU04] X. Liu, A.J. Goldsmith. Wireless Network Design for Distributed Control. *Proc. IEEE Conf. on Decision and Control*, 2004.

[LIU05a] H. Liu, S. Fei. Optimal Dynamic Scheduling Algorithm for Real-Time Tasks in Digital Control Systems. *Chinese Journal of Control and Decision*, Vol. 20, No.3, pp.246 – 250, 2005.

[LIU05b] L.P. Liu, F. Xia, Z. Wang, J.M. Chen, Y.X. Sun. Deployment Issues in Wireless Sensor Networks. *MSN'05*, *Lecture Notes in Computer Science*, Vol. 3794, pp.239 – 248, 2005.

[LIU06] X. Liu, J. Heo, L. Sha, X. Zhu. Adaptive Control of Multi-tiered Web Applications Using Queueing Predictor. *Proc. 2006 IEEE/IFIP Network Operations & Management Symposium*, Vancouver, Canada, 2006.

[LOY03] J.P. Loyall. Emerging Trends in Adaptive Middleware and Its Application to Distributed Real-Time Embedded Systems. *Lecture Notes in Computer Science*, Vol. 2855, pp.20 – 34, 2003.

[LU99] C. Lu., J. Stankovic, G. Tao, S.H. Son. Design and Evaluation of a Feedback Contol EDF Scheduling Algorithm. *Proc. 20th IEEE RTSS*, pp.56 – 67, 1999.

[LU00] C. Lu, J. Stankovic, T. Abdelzaher, G. Tao, S. Son, M. Marley. Performance Specifications and Metrics for Adaptive Real-Time Systems. *Proc. 21st IEEE RTSS*, pp.13 – 23, 2000.

[LU01] Y. Lu, A. Saxena, T.F. Abdelzaher. Differentiated Caching Services: A Control-theoretic Approach. *Proc. IEEE ICDCS*, 2001.

[LU02a] C. Lu, J.A. Stankovic, G. Tao, S.H. Son. Feedback Control Real-Time Scheduling: Framework, Modeling, and Algorithms. *Real-Time Systems*, Vol.23, No.1/2, pp.85 – 126, 2002.

[LU02b] Z. Lu, J. Hein, M. Humphrey, M. Stan, J. Lach, K. Skadron. Control-Theoretic Dynamic Frequency and Voltage Scaling for Multimedia Workloads. *Proc. Int'l Conf. Compilers, Architectures, & Synthesis Embedded Syst. (CASES)*, pp.156 – 163, 2002.

[LU03] Z. Lu, J. Lach, M.R. Stan, K. Skadron. Reducing Multimedia Decode Power using Feedback Control. *Proc. 21st Int. Conf. Computer Design*, pp.489 – 496, 2003.

[LU05] C. Lu, X. Wang, X. Koutsoukos. Feedback Utilization Control in Distributed Real-Time Systems with End-to-End Tasks. *IEEE Transactions on Parallel and Distributed Systems*, Vol.16, No.6, pp. 550 – 561, 2005.

[MAR02a] P. Martí. Analysis and Design of Real-Time Control Systems with Varying Control Timing Constraints. Ph.D Thesis, Technical University of Catalonia, 2002.

[MAR02b] P. Marti, G. Fohler, K. Ramamritham, J.M. Fuertes. Improving Quality-of-control Using Flexible Time Constraints: Metric and Scheduling Issues. *Proc. 23rd IEEE RTSS*, Austin, TX, USA, pp.91 – 100, 2002.

[MAR04] P. Martí, C. Lin, S. Brandt, M. Velasco, J.M. Fuertes. Optimal State Feedback Based Resource Allocation for Resource-Constrained Control Tasks. *25th IEEE Real-Time Systems Symposium (RTSS04)*, Lisbon, Portugal, 2004.

[MAT05] M. Mathiesen, G. Thonet, N. Aakwaag. Wireless Ad-hoc Networks for Industrial Automation: Current Trends and Future Prospects. *Proc. of the IFAC World Congress*, Prague, Czech Republic, 2005.

[MOS04] Y. Mostofi, R. Murray. Effect of Time-Varying Fading Channels on the Control Performance of a Mobile Sensor Node. *Proc. of IEEE 1st Int. Conf. on Sensor and Ad Hoc Communications and Networks (Secon)*, Santa Clara, CA, 2004.

[MUR03] R.M. Murray (Editor). Control in an Information Rich World: Report of the Panel on Future Directions in Control, Dynamics, and Systems. *Society for Industrial and Applied Mathematic*, 2003.

[NIK05] G. Nikolakopoulos, A. Panousopoulou, A. Tzes, J. Lygeros. Multihoping Induced Gain Scheduling for Wireless Networked Controlled Systems. *Proc. of the 44th IEEE Conf. on Decision and Control, and the European Control Conf.*, Seville, Spain, 2005.

[NIL98] J. Nilsson. Real-Time Control Systems with Delays. Ph.D Thesis, Lund University, 1998.

[PAL02a] L. Palopoli. Design of Embedded Control Systems under Real-time Scheduling Constraints. Ph. D Thesis, Scuola Superiore S. Anna, 2002.

[PAL02b] L. Palopoli, C. Pinello, A. S. Vincentelli, L. Elghaoui, A. Bicchi. Synthesis of Robust Control Systems under Resource Constraints. *Lecture Notes in Computer Science (HSCC'02)*, Stanford, USA, 2002.

[PAL02c] L. Palopoli, L. Abeni, G. Bolognini, B. Allotta, F. Conticelli. Novel Scheduling Policies in Real-Time Multithread Control System Design. *Control Engineering Practice*, Vol. 10, pp.1095 – 1110, 2002.

[PAL03] L. Palopoli, L. Abeni, G. Lipari. On the applications of Hybrid Control to CPU Reservations. HSCC'03, *Lecture Notes in Computer Science*, Vol. 2623, pp.389 – 404, 2003.

[PAN04] A. Panousopoulou, G. Nikolakopoulos, A. Tzes. Experimental Controller Tuning and QoS Optimization of a Wireless Transmission Scheme for Real-Time Remote Control Applications. *Proc. of the 2004 IEEE Int. Conf. on Industrial Technology*, 2004.

[PAR02] H. S. Park, Y.H. Kim, D.S. Kim, W.H. Kwon. A Scheduling Method for Network-based Control Systems. *IEEE Translations on Control Systems Technology*, Vol. 10, No.3, pp.318 – 330, 2002.

[PIL01] P. Pillai, K.G. Shin. Real-Time Dynamic Voltage Scaling for Low Power Embedded Operating Systems. *Proc. 18th ACM Symposium on Operating Systems Principles*, Banff, Alberta, Canada, pp.89 – 102, 2001.

[PLO04] N. Ploplys, P. Kawka, A. Alleyne. Closed Loop Control over Wireless Networks. *IEEE Control Systems Magazine*, pp.58 – 71, 2004.

[PRI05] F.D. Priscoli, A. Isidori. A Control Engineering Approach to Integrated Congestion Control and Scheduling in Wireless Local Area Networks. *Control Engineering Practice*, Vol. 13, pp. 541 – 558, 2005.

[RAM05] H. Ramamurthy, D. Lai, B. S. Prabhu, R. Gadh. ReWINS: A Distributed Multi-RF Sensor Control Network for Industrial Automation. *IEEE Wireless Telecommunication Symposium WTS*, Pomona, California, pp.24 – 33, 2005.

[RAV01] B. Ravindran, P. Kachroo, T. Hegazy. Intelligent Feedback Control-based Adaptive Resource Management for Asynchronous. *Decentralized Real-Time Systems*, *IEEE Trans. Sys., Man, and Cybernetics*, Part C, Vol. 31, No. 02, pp.261 – 265, 2001.

[RAY88] A. Ray, Y. Halevi. Integrated Communication and Control Systems: Part II—Design Considerations. *ASME Journal of Dynamic Systems, Measurement and Control*, Vol. 110, pp.374 – 381, 1988.

[REN02] F. Ren, Y. Ren, X. Shan. Design of a Fuzzy Controller for Active Queue Management. *Computer Communications*, Vol. 25, No.9, pp. 874 – 883, 2002.

[REN03] X.D. Ren, S.B. Li, Z. Wang, Y.X. Sun. A QoS Management Scheme for Paralleled Networked Control Systems with CAN Bus. *Proc. 29th IEEE IECON*, pp.842 – 847, 2003.

[REH00] H. Rehbinder, M. Sanfridson. Integration of Offline Scheduling and Optimal Control. *Proc. 12th ECRTS*, Sweden, 2000.

[ROB05] D. Robert, O. Sename, D. Simon. Sampling Period Dependent RST Controller Used in Control/Scheduling Codesign. *Proc. IFAC World Congress*, Czech Republic, 2005.

[RYU97] M. Ryu, S. Hong, M. Saksena. Streamlining Real-Time Controller Design: From Performance Specifications to End-to-end Timing Constraints. *Proc. IEEE RTAS*, Montreol, Canada, pp.91 – 99, 1997.

[SAH02] D.R. Sahoo, S. Swaminathan, R. Al-Omari, M.V. Salapaka, G. Manimaran, A.K. Somani. Feedback Control for Real-Time Scheduling. *Proc. IEEE ACC*, Vol.2, pp.1254 – 1259, 2002.

[SAN04] M. Sanfridson. Quality of Control and Real-Time Scheduling, Ph.D Thesis. Royal Institute of Technology (KTH), 2004.

[SAN05] R. Sandoval-Rodriguez, C.T. Abdallah, H.N. Jerez, R.H. Byrne. Experimental Results on the Effects of 802.11b WLAN on Networked Control System. *13th Mediterranean Conf. on Control and Automation*, Limassol, Cyprus, pp.1113 – 1118, 2005.

[SET96] D. Seto, J.P. Lehoczky, L. Sha, K.G. Shin. On Task Schedulability in Real-Time Control Systems. *Proc. 17th IEEE RTSS*, Washington, DC, pp.13 – 21, 1996.

[SET98] D. Seto, J.P. Lehoczky, L. Sha. Task Period Selection and Schedulability in Real-Time Systems. *Proc. of the 19th IEEE Real-Time Systems Symposium*, *Madrid*, pp.188 – 198, 1998.

[SHA48] C.E. Shannon. A Mathematical Theory of Communication. *Bell System Technical Journal*, Vol.27, No.4, pp.623 – 656, 1948.

[SHA00] L. Sha, X. Liu, M. Caccamo, G. Buttazzo. Online Control Optimization Using Load Driven Scheduling. *Proc. of the 39th Conf. on Decision and Control*, *Sydney*, Vol.5, pp.4877 – 4882, 2000.

[SHA02] L. Sha, X. Liu, Y. Lu, T. Abdelzaher. Queueing Model Based Network Server Performance Control. *IEEE RTSS*, pp.81 – 90, 2002.

[SHA04] L. Sha, T. Abdelzaher, K. E. Årzén, T. Baker, A. Burns, G. Buttazzo, M. Caccamo, A. Cervin, J. Lehoczky, A. Mok. Real-Time Scheduling Theory: A Historical Perspective. *Real-Time Systems*, Vol.28, pp.101 – 155, 2004.

[SIM05] D. Simon, D. Robert, O. Sename. Robust Control/Scheduling Co-Design: Application to Robot Control. *Proc. IEEE RTAS*, California, USA, 2005.

[SON05] I. Song, S. Kim, F. Karray. Stability Analysis of Feedback Controlled Reservation-based CPU Scheduler. *1st Int. ECRTS Workshop on Real-Time and Control*, Palma de Mallorca, Spain, 2005.

[SOR05] A. Soria-Lopez P. Mejia-Alvarez, J. Cornejo. Feedback Scheduling of Power-Aware Soft Real-Time Tasks. *Proc. of the Sixth Mexican Int. Conf. on Computer Science (ENC'05)*, pp.266 – 273, 2005.

[STA96] J.A. Stankovic. Real-Time and Embedded Systems. *ACM Computing Surveys*, Vol. 28, No.1, pp.205 – 208, 1996.

[STA99] J.A. Stankovic, C. Lu, S.H. Son, G. Tao. The Case for Feedback Control Real-Time Scheduling. *Proc. 11th ECRTS*, York, UK, pp. 11 – 20, 1999.

[STA01] J.A. Stankovic, T. He, T.F. Abdelzaher, M. Marley, G. Tao, S.H. Son, C. Lu. Feedback Control Real-Time Scheduling in Distributed Real-Time Systems. *IEEE Real-Time Systems Symposium (RTSS 2001)*, pp.59 – 70, 2001.

[STE00] D. Steere, M.H. Shor, A. Goel, J. Walpole, C. Pu. Control and Modeling Issues in Computer Operating Systems: Resource Management for Real-rate Computer Applications. *39th IEEE Conf. on Decision and Control (CDC)*, 2000.

[TAT04] S. Tatikonda, S. Mitter. Control Under Communication Constraints. *IEEE Transactions on Automatic Control*, Vol.49, No.7, pp.1056 – 1068, 2004.

[THO05] J. P. Thomesse. Fieldbus Technology in Industrial Automation. *Proceedings of the IEEE*, Vol.93, No.6, pp.1073 – 1101, 2005.

[TIA06] Y.C. Tian, Q.L. Han, D. Levy, M.O. Tadé. Reducing Control Latency and Jitter in Real-Time Control. *Asian Journal of Control*, Vol.8, No.1, pp.72 – 75, 2006.

[TIP04] Y. Tipsuwan, M.Y. Chow. Gain Scheduler Middleware: A Methodology to Enable Existing Controllers for Networked Control and Teleoperation—Part I: Networked Control. *IEEE Transactions on Industrial Electronics*, Vol. 51, No. 6, pp.1218 – 1227, 2004.

[TON04] L. Tong, X. Huai, M. Li. A Time-sharing Scheduling Algorithm Based on PID Feedback Control. *Chinese Journal of Computer Research and Development*, Vol. 40, No.1, pp.15 – 21, 2004.

[TOV99] E. Tovar. Supporting Real-Time Communications with Standard Factory-Floor Networks. Ph.D Thesis, University of Porto, 1999.

[TZE03] A. Tzes, G. Nikolakopoulos, I. Koutroulis. Development and Experimental Verification of a Mobile Client—Centric Networked Controlled System. *Proc. of the European Control Conf.* , London, 2003.

[UNS03] O.S. Unsal, I. Koren. System-level Power-aware Design Techniques in Real-Time Systems. *Proceedings of the IEEE*, Vol.91, No.7, pp. 1055 – 1069, 2003.

[VAR03] A. Varma, B. Ganesh, M. Sen, S.R. Choudhury, L. Srinivasan, J. Bruce. A Control-theoretic Approach to Dynamic Voltage Scheduling. *Proc. ACM CASES*, Georgia, USA, pp.255 – 266, 2003.

[VEL04] M. Velasco, J.M. Fuertes, C. Lin, P. Martí, S. Brandt. A Control Approach to Bandwidth Management in Networked Control Systems. *30th Annual Conf. of the IEEE Industrial Electronics Society (IECON04)*, Busan, Korea, 2004.

[WAL01] G.C. Walsh, H. Ye. Scheduling of Networked Control Systems. *IEEE Control Systems Magazine*, Vol. 21, No. 1, pp.57 – 65, 2001.

[WAN05] X. Wang, D. Jia, C. Lu, X. Koutsoukos. Decentralized Utilization Control in Distributed Real-Time Systems. *IEEE Real-Time Systems Symposium RTSS'05*, 2005.

[WAN06] H.A. Wang, H. Jin, H. Wang, G.Z. Dai. Energy-Aware Optimization Design of Digital Control Systems with Evolution Strategy. *Dynamics of Continuous, Discrete and Impulsive Systems-Series B: Applications and Algorithms, Special Issue on ICSCA'06*, Watam Press, pp.1893 – 1898, 2006.

[WEI94] M. Weiser, B. Welch, A. Demers, S. Shenker. Scheduling for Reduced CPU Energy. *First USENIX Symposium on Operating Systems Design and Implementation*, Montreal, Canada, pp.13 – 23, 1994.

[WEI02] L. Wei, W. Ma, H. Yu. Study on Sampling Period in Real-Time Scheduling of Feedback Control. *Chinese Journal of Process Automation Instrumentation*, Vol. 23, No.9, pp.25 – 29, 2002.

[WEI03] L. Wei, H. Yu. Research on a Soft Real-Time Scheduling Algorithm Based on Hybrid Adaptive Control Architecture. *Proc. IEEE ACC*, Denver, USA, pp.4022 – 4027, 2003.

[WEI05] J. Wei, C. Xu. A Self-Tuning Fuzzy Control Approach for End-to-End QoS Guarantees in Web Servers. *Proc. Int'l Workshop Quality of Service, Lecture Notes in Computer Science*, Vol. 3552, pp.123 – 135, 2005.

[WIE48] N. Wiener. *Cybernetics, or Control and Communication in the Animal and the Machine.* Cambridge, MA: The MIT Press, 1948.

[WIL05] A. Willig, K. Matheus, A. Wolisz. Wireless Technology in Industrial Networks. *Proceedings of the IEEE*, Vol.93, No.6, pp.1130 – 1151, 2005.

[XIA03] L. Xiao, M. Johansson, H. Hindi, S.Boyd, A. Goldsmith. Joint Optimization of Communication Rates and Linear Systems. *IEEE Trans. Automatic Control*, Vol. 48, No. 1, pp.148 – 153, 2003.

[XIA04a] F. Xia, Z. Wang, Y.X. Sun. Simulation Based Performance Analysis of Networked Control Systems with Resource Constraints. *Proc. of 30th IEEE IECON*, Busan, Korea, 2004.

[XIA04b] F. Xia, Z. Wang, Y.X. Sun. Integrated Computation, Communication and control: Towards Next Revolution in Information Technology. *Lecture Notes in Computer Science*, Vol. 3356, pp.117 – 125, 2004.

[XIA05a] F. Xia, S.B. Li, Y.X. Sun. Neural Network Based Feedback Scheduler for Networked Control System with Flexible Workload. *Lecture Notes in Computer Science*, Vol. 3611, pp.237 – 246, 2005.

[XIA05b] F. Xia, Y.X. Sun. NN-based Iterative Learning Control under Resource Constraints: A Feedback Scheduling Approach. *Lecture Notes in Computer Science*, Vol.3498, pp.1 – 6, 2005.

[XIA05c] F. Xia, Y.X. Sun. Anytime Iterative Optimal Control Using Fuzzy Feedback Scheduler. *Lecture Notes in Artificial Intelligence*, Vol. 3682, pp.350 – 356, 2005.

[XIA06a] F. Xia, X.H. Dai, Y.X. Sun, J.X. Shou. Control Oriented Direct Feedback Scheduling. *International Journal of Information Technology*, Vol. 12, No.3, pp.21 – 32, 2006.

[XIA06b] F. Xia, Y.X. Sun. Control-Scheduling Codesign: A Perspective on Integrating Control and Computing. *Dynamics of Continuous, Discrete and Impulsive Systems-Series B: Applications and Algorithms, Special Issue on ICSCA'06*, Watam Press, pp.1352 – 1358, 2006.

[XU06a] W.Q. Xu. Study on Congestion Control in Ad Hoc Networks. Ph.D Thesis, Zhejiang University, 2006.

[XU06b] W. Xu, X. Zhu, S. Singhal, Z. Wang. Predictive Control for Dynamic Resource Allocation in Enterprise Data Centers. *Proc. IEEE/IFIP Network Operations & Management Symposium*, Vancouver, Canada, pp.115 – 126, 2006.

[YE01] H. Ye, G.C. Walsh, L. Bushnell. Real-Time Mixed Traffic Wireless Networks. *IEEE Transactions on Industrial Electronics*, Vol. 48, No. 5, pp.883 – 890, 2001.

[YEP02] J. Yépez, P. Martí, J.M. Fuertes. Control Loop Performance Analysis over Networked Control Systems. *28th Annual Conf. of the IEEE Industrial Electronics Society (IECON02)*, Sevilla, Spain, Vol.4, pp. 2880 – 2885, 2002.

[YEP03] J. Yépez, P. Martí, J.M. Fuertes. The Large Error First (LEF) Scheduling Policy for Real-Time Control Systems. *Work-in-Progress Session of the 24th IEEE Real-Time Systems Symposium (RTSS03)*, Cancun, Mexico, 2003.

[YIN04] H.X. Yin, F. Xia, Z. Wang, Y.X. Sun. Performance Evaluation of Hard Real-Time Data in the Switched Ethernet by Network Calculus. *Proc. 5th World Congress on Intelligent Control and Automation*, Hangzhou, China, Vol.2, pp.1451 – 1455, 2004.

[YUA05] P. Yuan, M. Moallem, R.V. Patel. A Feedback Scheduling Algorithm for Real Time Control Systems. *Proc. of the 2005 IEEE Conf. on Control Applications*, Toronto, Canada, August 28 – 31, pp.873 – 878, 2005.

[ZHA01] W. Zhang. Stability Analysis of Networked Control Systems. Ph.D Thesis, Department of Electrical Engineering and Computer Science, Case Western Reserve University, 2001.

[ZHA04] Y. Zhang, C. Fang, Y. Wang. A Feedback-Driven Online Scheduler for Processes with Imprecise Computing. *Chinese Journal of Software*, Vol. 15, No.4, pp.616 – 623, 2004.

[ZHA06a] W.H. Zhao, F. Xia. Dynamic Voltage Scaling with Asynchronous Period Adjustment for Embedded Controllers. *Dynamics of Continuous, Discrete and Impulsive Systems-Series B: Applications and Algorithms, Special Issue on ICSCA'06*, pp.514 – 519, Watam Press, 2006.

[ZHA06b] W.H. Zhao, F. Xia. An Efficient Approach to Energy Saving in Microcontrollers. *Lecture Notes in Computer Science*, Vol. 4186, pp.595 – 601, 2006.

[ZHA06c] W.H. Zhao, F. Xia. A Neural Network Approach to QoS Management in Networked Control Systems over Ethernet. *Lecture Notes in Control and Information Sciences*, Vol. 344, pp.444 – 449, 2006.

[ZHO05a] P. Zhou, J. Xie. Online Deadline Assignment to Reduce Output Jitter of Real-Time Control Tasks. *Proc. of Int. Conf. on Machine Learning and Cybernetics*, Vol. 2, pp.1312 – 1315, 2005.

[ZHO05b] P. Zhou, J. Xie, X. Deng. Feedback Scheduling to Optimize the Performance of Embedded Real-Time Control Systems. *Journal of Information & Computational Science*, Vol. 2, No.3, pp.617 – 623, 2005.

[ZHU05a] Y. Zhu, F. Mueller. Feedback EDF Scheduling of Real-Time Tasks Exploiting Dynamic Voltage Scaling. *Real-Time Systems*, Vol.31, No. 1 – 3, pp.33 – 63, 2005.

[ZHU05b] Y. Zhu. Dynamic Voltage Scaling with Feedback EDF Scheduling for Real-Time Embedded Systems. Ph.D Thesis, North Carolina State University, 2005.

Chapter 2

Introduction to Feedback Scheduling

Feedback scheduling is a promising technology for dealing with uncertainty in resource availability in Real-Time Control Systems. This chapter begins with a brief overview of fundamental yet important concepts and principles of sampled-data control and real-time scheduling, providing required background knowledge for understanding the methodology of feedback scheduling. The temporal attributes of control loops including delay and its jitter, sampling period and its jitter, and data loss are described. Methods used in this book to handle these temporal attributes are also given. A concrete case is studied via simulations to demonstrate how resource scheduling impacts the control performance of concurrent multiple loops, thus illustrating the necessity of applying feedback scheduling. Some important concepts related to feedback scheduling are explained. Together with a basic framework, some critical design issues on feedback scheduling are discussed, thus constituting a unified platform for the work to be reported in subsequent chapters.

2.1 Fundamentals of Sampled-Data Control

The concept of *feedback* is the central element of control, a discipline that deals with the regulation of the characteristics of a system (i.e. physical process). The main idea of feedback control [AST97, HEL04] is to exploit measurements of the system's outputs to determine the control commands that yield the desired system behaviour. A feedback controller, together with some sensors and actuators, is usually used to sense the operation of a system, compare it against the desired behaviour, compute control commands, and actuate the system to effect the desired change.

Over the past 40 years, the field of control has provided a huge amount of principles and methods that can be used to design systems maintaining desirable performance by automatically adapting to changes in the system and/or the environment. With these principles and methods, the system can be driven to respond properly even if its dynamic behaviour is not exactly known or if external disturbances that tend to cause it to respond incorrectly are present. In this way, feedback control has been serving many fields as a means of ensuring robustness to uncertainty. This explains why feedback control systems are now all around us in the modern technological world [MUR03].

In control terms, the *controlled variable* (i.e. *measured output*) is a measurable characteristic of the system to be controlled; the *manipulated variable* (i.e. *control input*) is a parameter that can be dynamically adjusted to affect the controlled variable of the system; the desired value of the controlled variable is referred to as *setpoint* or *reference input*.

The feedback architecture of a control system is also called *closed loop*, since the measured outputs are used to determine the control inputs, which in turn affect the system outputs. This motivates the use of the term *closed loop control* to refer to feedback control, which is in contrast to *open-loop control* (sometimes referred to as *feedforward control*), a technique that does not use feedback. An open-loop controller uses the setpoint to determine the setting of the manipulated variable needed to achieve the desired system response.

Although open-loop control has advantages in terms of reducing design complexity by, e.g. avoiding online measurements of the system outputs, there are numerous reasons why closed loop control is often preferable to open-loop control [HEL04, MAR02]. For example, open-loop control cannot compensate for disturbances or noise, whereas closed loop control can do this; open-loop control relies highly on an accurate model of the target system, something that is difficult to obtain in practice, whereas closed loop control does not require such an accurate system model. As a consequence, open-loop systems are rarely used in practice.

2.1.1 Architecture

A schematic diagram of a typical real-time control loop is given in Fig.2.1. The main components consist of the controlled process, a sensor that contains an A/D (Analog-to-Digital) converter, an embedded computer/controller, an actuator that contains a D/A (Digital-to-Analog) converter, and a communication network.

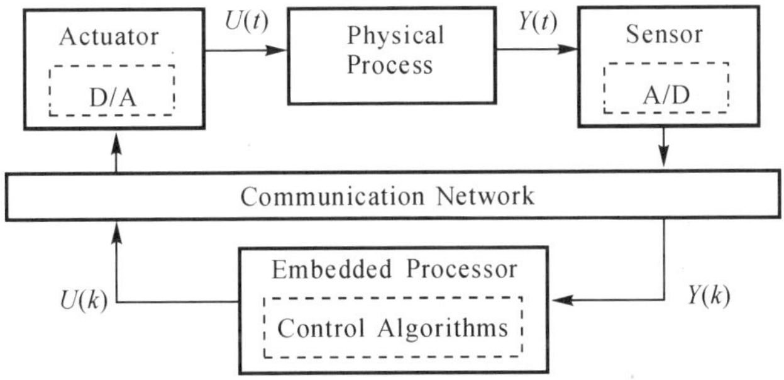

Fig. 2.1 Schematic diagram of a real-time control loop

The controlled system is generally a continuous-time physical proccss, e.g. DC motor, inverted pendulum, etc. The inputs and outputs of the process are continuous-time signals. The A/D converter transforms the outputs of the process into digital signals at *sampling instants*. The time between two consecutive sampling instants is the *sampling period* and denoted by h. The A/D converter could be either a separated unit, or embedded into the sensor. The controller takes charge of executing software programs that process the sequence of sampled data according to specific control algorithms and then produce the sequence of control commands. To make these digital signals applicable to the physical process, the D/A converter transforms them into continuous-time signals with the help of a hold circuit that determines the input to the process until a new control command is available from the controller. The most common method is the zero-order-hold that holds the input constant over the sampling period. Another type of hold device is the first-order-hold, where the input is computed from current and previous outputs from the controller using a first order polynomial in time. In a networked environment, the sequences of sampled data and the control commands need to be transmitted from the sensor to the controller and from the controller to the actuator, respectively, over the communication network. The network could either be wired (e.g. fieldbuses and Internet) or be wireless (e.g. WLAN, ZigBee, and Bluetooth).

Generally speaking, every unit in the system may be either time-trigged or event-triggered. Unless otherwise specified, in this book we consider systems where the sensor is time-triggered, while the controller, the actuator, and the communication network are event-triggered.

In a control loop shown in Fig.2.1, the operations including A/D conversion, execution of control programs, D/A conversion, and data transmission undoubtedly need a certain amount of time to accomplish. When the computing and communication resources (e.g. the CPU time and network bandwidth) are limited, there must be a significant time delay from the sampling of the sensor to the actuation of the corresponding control command. As mentioned in Chapter 1, the properties of timing parameters associated with real-world control loops may change due to many factors such as resource sharing and workload variations, which affects in turn the QoC of control systems.

An instance of the timing diagram of a control loop is given in Fig.2.2. Herein some phenomena have been observed: 1) delay is present, 2) delay may be time-varying, 3) delay may be longer than the sampling period, and 4) data may be lost.

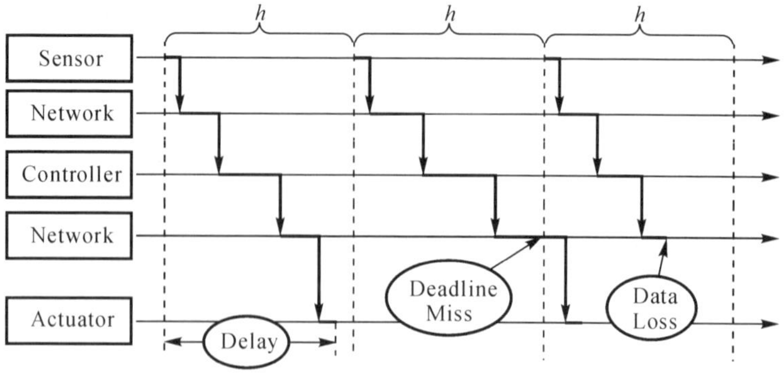

Fig. 2. 2 Timing diagram of a control loop

These phenomena will be further discussed in Section 2.3.

2.1.2 Design Methods

Sampled-data control systems are inherently a class of hybrid systems that contain both continuous-time and discrete-time signals. There are two different approaches for the design of sampled-data control systems [ARZ99, AST97, WIT02]: discrete-time (direct) design and discretization of a continuous-time design.

Discrete-Time Design

The first design method looks upon the process from the view of the computer. Only the values of the system inputs and outputs at sampling instants will be considered. Treating the process as a discrete-time system, the digital controller can be directly designed in the discrete-time domain. With this method, the physical process is still a continuous-time system and any information about what is happening between the sampling instants is disregarded.

To design the controller, a sampled version of the system model must be derived first. Assume that periodic sampling and zero-order-hold are used. For a continuous-time system that is described by the time-invariant state-space model:

$$\begin{aligned} \frac{\mathrm{d}x}{\mathrm{d}t} &= Ax(t) + Bu(t) \\ y(t) &= Cx(t) + Du(t) \end{aligned}, \tag{2.1}$$

solving the system equation gives the system state on the interval $kh \leqslant t < kh + h$ as follows:

$$\begin{aligned} x(t) &= \mathrm{e}^{A(t-kh)}x(kh) + \int_{kh}^{t} \mathrm{e}^{A(t-s')}Bu(s')\mathrm{d}s' \\ &= \mathrm{e}^{A(t-kh)}x(kh) + \int_{kh}^{t} \mathrm{e}^{A(t-s')}\mathrm{d}s'Bu(kh) \\ &= \mathrm{e}^{A(t-kh)}x(kh) + \int_{0}^{t-kh} \mathrm{e}^{As}\mathrm{d}sBu(kh) \\ &= \Phi(t, kh)x(kh) + \Gamma(t, kh)u(kh) \end{aligned}. \tag{2.2}$$

The equation above describes the change of the state over the sampling period. By considering the situation at the next sampling instant, the following discrete-time system description can be obtained:

$$\begin{aligned} x(kh+h) &= \Phi x(kh) + \Gamma u(kh) \\ y(kh) &= Cx(kh) + Du(kh) \end{aligned}, \tag{2.3}$$

where

$$\begin{aligned} \Phi &= \mathrm{e}^{Ah} \\ \Gamma &= \int_{0}^{h} \mathrm{e}^{As}\mathrm{d}sB \end{aligned}. \tag{2.4}$$

Using similar techniques we can derive the sampled version of continuous-time system model that is also valid for aperiodic sampling. With this accomplished, a wide range of discrete-time controller design methods such as pole placement design and linear quadratic design can then be applied.

Discretization of a Continuous-Time Design

Another way to design a sampled-data control system is to make a continuous-time design and then make a discrete-time approximation of this controller by discretization. The sampled-data control system designed in this way should behave as the continuous-time system provided that the sampling period is sufficiently small. Furthermore, no discrete-time control theory is required for this approach.

When the continuous-time controller is already available and specified as a transfer function $G(s)$, the goal is then to look for methods that will transform $G(s)$ to a pulse-transfer function $H(z)$ such that A/D + $H(z)$ + D/A $\approx G(s)$. This can be done in several ways.

The most straightforward way is to use the forward approximation (also known as Euler's method) or the backward approximation. In the forward approximation a derivative is replaced by its forward approximation, i.e.

$$\frac{\mathrm{d}x}{\mathrm{d}t} \approx \frac{x(t+h) - x(t)}{h} \Leftrightarrow z = \mathrm{e}^{sh} \approx 1 + sh \ .$$

In the backward approximation the derivative is replaced by:

$$\frac{\mathrm{d}x}{\mathrm{d}t} \approx \frac{x(t) - x(t-h)}{h} \Leftrightarrow z = \mathrm{e}^{sh} \approx \frac{1}{1 - sh} \ .$$

A third method is Tustin's approximation given by:

$$z = \mathrm{e}^{sh} \approx \frac{1 + sh/2}{1 - sh/2} \ .$$

Using the approximation methods above, $H(z)$ can be obtained simply by replacing s in $G(s)$ by s', i.e. $H(z) = G(s')$, where

$$s' = \begin{cases} \dfrac{z-1}{h} & \text{(Forward approximation)} \\ \dfrac{z-1}{zh} & \text{(Backward approximation)} \ . \\ \dfrac{2}{h} \cdot \dfrac{z-1}{z+1} & \text{(Tustin's method)} \end{cases} \tag{2.5}$$

The relative merits of different approximation methods have been discussed in [WIT02]. When the forward approximation is used, a stable continuous-time system may be transformed into an unstable discrete-time system. On the contrary, a stable continuous-time system will always give a stable discrete-time system with the backward approximation. However, it is possible that an unstable continuous-time system is mapped into a stable discrete-time system in the case of backward approximation. Tustin's method is advantageous in the sense that the left half s-plane is exactly mapped into the unit disc on the z-plane. Stable continuous-time systems are always transformed into stable discrete-time systems and unstable discrete-time systems always corresponds to unstable continuous-time systems.

2.1.3 Quality of Control

The most fundamental requirement of a control system is stability. In the control community, there are various versions of the definition of stability. For example, a system is said to be Bounded-Input Bounded-Output (BIBO) *stable* if for any bounded input, the output is bounded. If the system output neither converges nor becomes unbounded, the system is said to be *marginally stable*. A system is *unstable* if the system output unboundedly diverges. The stability of a closed loop system can be determined by the location of its closed loop poles. A linear time-invariant control system is stable when in the continuous-time domain all its closed loop poles are in the left half s-plane, or in discrete-time domain all its closed loop poles are within the unit circle on the z-plane.

Apart from stability, the transient behaviour is another focus of attention for control systems design. There are several properties that can be used to evaluate the transient behaviour of a closed loop control system, for example, *steady-state accuracy*, *settling time*, and *overshoot*.

Beyond these properties, more widely used criteria for QoC relate to the *control error* $e(t)$, which is defined as the difference between the setpoint $r(t)$ and the system output $y(t)$. Some of these performance indices are given below in both continuous-time and discrete-time forms, where t_0 (k_0) and t_f (k_f) are the initial and final continuous (discrete) times of the evaluation period.

Integral of Absolute Error (IAE):

$$IAE = \int_{t_0}^{t_f} |e(t)| \,\mathrm{d}t \approx \sum_{k=k_0}^{k_f} |r(kh) - y(kh)|. \tag{2.6}$$

Integral of Time-weighted Absolute Error (ITAE):

$$ITAE = \int_{t_0}^{t_f} t|e(t)| \,\mathrm{d}t \approx \sum_{k=k_0}^{k_f} k|r(kh) - y(kh)|. \tag{2.7}$$

Integral of Square Error (ISE):

$$ISE = \int_{t_0}^{t_f} e^2(t) \,\mathrm{d}t \approx \sum_{k=k_0}^{k_f} (r(kh) - y(kh))^2. \tag{2.8}$$

Integral of Time-weighted Square Error (ITSE):

$$ITSE = \int_{t_0}^{t_f} te^2(t) \,\mathrm{d}t \approx \sum_{k=k_0}^{k_f} k(r(kh) - y(kh))^2. \tag{2.9}$$

While IAE and ISE weight all errors equally, ITAE and ITSE weight errors with time penalizing later errors more heavily. Different criteria could be chosen to meet different requirements of performance evaluation. For any of the criteria above, the bigger the value the worse the QoC. From this insight, the value of a QoC index is also referred to as *control cost*.

2.2 Scheduling in Real-Time Systems

A *real-time system* is a system in which the correctness of the system depends not only on the logical results of the computation, but also on the time at which the results are produced [LIU00, STA96]. Real-Time systems have to respond to external events within a finite and specified period of time. Nevertheless, real-time does not necessarily mean fast computation or communication. Rather, the key in real-time systems is predictability, implying that the system should execute at a speed that meets the timing requirements of the external physical world.

It is common to divide real-time systems into two types: hard and soft. In a *hard real-time system* it is absolutely imperative that the response finishes within the specified deadline. Many hard real-time systems perform safety-critical tasks where catastrophic damage or even death is at risk if the deadlines are not met. Examples of hard real-time systems include flight control systems, vehicle brake-by-wire systems, and critical information systems. In a *soft real-time system*, on the contrary, the deadlines are important but the system will still function correctly in the face of occasional deadline misses. Examples of soft real-time systems include Web servers, multimedia, and telephone switches.

A real-time system may contain different types of resources including both computing and communication resources. The problem of resource scheduling in real-time systems can be informally formulated as follows [PAL02]: *Given a set of timing constraints for the computing tasks or for the communication messages, find a schedule of the shared resources such that all constraints are met.* Depending on the type of the resources to be allocated, scheduling in real-time systems can be classified into two branches: real-time computing and real-time communication. In the rest of this section, we outline some important concepts and techniques in real-time scheduling of processors and networks, respectively. For further readings see for example [LIU00, SHA04].

2.2.1 Real-Time Computing

In real-time computing systems, the processor (i.e. CPU) is often shared by multiple tasks. In this context a *task* is a software program consisting of a sequence of instructions that are intended to be repeatedly executed on the processor. Each activation of a task corresponds to the execution of a *job* (or *task instance*). Each job may have its release time, start time, preemption time, resumption time, and completion time. At a minimum, a real-time task i is characterized by a *Worst-Case Execution Time* (WCET) C_i and a *relative deadline* D_i. The WCET is the worst-case CPU time it takes to execute the sequence of instructions associated with the task in the absence of other competing tasks. The relative deadline is the time interval between the release time of the job and its *absolute deadline*, the time by which the job has to terminate.

A task can be periodic, aperiodic, or sporadic, depending on how the corresponding jobs are activated. A task that is activated every T_i clock unit is called a *periodic task*. The (relative) deadline for each activated instance may be less than, equal to, or larger than the period T_i. A task that is activated at unpredictable times is an *aperiodic task*. A *sporadic task* is an aperiodic task subjected to the additional constraint that there is a minimum inter-arrival time between task activations.

Additional constraints besides timing constraints may apply to real-time tasks. For example, the tasks are said to have precedence constraints if they are constrained to execute in a specific order. Otherwise, if the tasks do not communicate or share resources other than the processor or network, implying that they can execute in any order, they are said to be *independent*. In this book only independent and periodic tasks will be considered.

Real-Time scheduling is the process of creating start and completion times for a set of tasks in such a way that all constraints are met. Tasks are generally scheduled based on a scheduling algorithm. A scheduler is the module that implements this algorithm. It produces schedules that assign all the tasks on the available resources. A schedule is feasible if every job completes by its deadline. A set of tasks are *schedulable* according to a scheduling algorithm if the scheduler always produces a feasible schedule when using the algorithm.

Real-Time CPU scheduling algorithms can be either *offline* or *online*. Offline scheduling takes advantage of *a priori* knowledge about the task parameters such as deadlines, execution times, precedence constraints, etc. A schedule of the tasks is generated offline and stored in a table for use at runtime. The scheduler schedules the jobs according to this schedule at each scheduling decision time.

Some advantages of using offline scheduling are: 1) the execution is very deterministic, 2) the scheduling overhead during runtime is minimized, and 3) complex dependencies can be handled. There are also many disadvantages with offline scheduling [LOK92]. For instance, it is not suitable for sporadic or aperiodic tasks; runtime flexibility is lacking in that a small adjustment of the system may possibly make it necessary to totally reconstruct the schedule; if a job happens to execute longer than scheduled it may cause the whole schedule to collapse. For these reasons, it is often preferable to use online scheduling [CER00, MAR02, WIT02].

Online scheduling algorithms make the decision on which task to execute at runtime. Online scheduling does not depend on *a priori* knowledge about the task set. It is much more flexible and adaptive than offline scheduling, though the scheduling overhead is often increased and the temporal predictability is reduced to some degrees. Particularly, in a system whose workload is unpredictable, online scheduling is the only option. An online scheduler can accommodate dynamic variations in user demands and resource availability.

For online scheduling, it is possible to distinguish between *preemptive scheduling* and *non-preemptive scheduling*. In preemptive scheduling, the execution of a job can be suspended because of the arrival of a job with higher priority, and resumed later. On the contrary, non-preemptive scheduling does not allow a running job to be interrupted. Once started, the execution of a job will continue until its completion.

Most preemptive scheduling algorithms are priority driven, which can be further classified into *fixed-priority scheduling* and *dynamic-priority scheduling*. In the following we outline the key results in these areas that relate to this work. Assumption is made that there is no kernel overhead in that context switches and interrupts take zero time. Non-preemptive scheduling will be discussed in Chapter 7.

Fixed-Priority Scheduling

A fixed-priority scheduling algorithm assigns the same priority to all the jobs of each task. This means that the priority of each task is fixed relative to other tasks. If several tasks are ready to run at the same time, the task with the highest priority gets access to the processor. Should a task with higher priority than the running task become ready, the running task is then preempted by the other task.

The most widely utilized fixed-priority algorithm is the Rate-Monotonic (RM) algorithm presented in the seminal work by Liu and Layland [LIU73]. This algorithm assigns priorities to tasks based on their periods: the shorter the period, the higher the priority. It has been shown that the RM algorithm is optimal when $D_i = T_i$ for all tasks. A fixed-priority (or dynamic-priority) scheduling algorithm is optimal in the sense that if the task set is not schedulable under this algorithm it will not be schedulable under any other fixed-priority (or dynamic-priority) scheduling algorithms.

A sufficient schedulability condition for RM scheduling is:

Theorem 2.1 A system that contains N independent, preemptive, periodic tasks with relative deadlines equal to periods (i.e. $D_i = T_i$) is schedulable according to the RM algorithm if the total utilization U satisfies that

$$U = \sum_{i=1}^{N} \frac{C_i}{T_i} \leqslant N(2^{\frac{1}{N}} - 1). \tag{2.10}$$

Since Theorem 2.1 gives only a sufficient condition, the system may still be schedulable if the CPU utilization is larger than the upper bound given by Eq.(2.10). For instance, Lehoczky, *et al.* [LEH89] showed that the average real feasible utilization for large randomly chosen task sets is about 88%. As the number of tasks approaches infinity, the utilization bound given by Eq.(2.10) approaches $\ln 2 \approx 0.693$. This leads to the following corollary:

Corollary 2.1 If the CPU utilization is no more than 69.3%, then the system described in Theorem 2.1 is schedulable with RM.

Necessary and sufficient feasibility tests have been performed for the RM algorithm based on e.g. calculation of the worst-case response time of each task. The *response time* of a task is defined as the time elapsed from its release to its completion. In accordance with the work by Joseph and Pandya [JOS86] and Audsley, *et al.* [AUD91], the following necessary and sufficient condition holds for

RM schedulability.

Theorem 2.2 The system described in Theorem 2.1 is RM schedulable if and only if $R_i \leqslant D_i$, $\forall i \in \{1, \cdots, N\}$, where R_i is the worst-case response time of task i, given by

$$R_i = C_i + \sum_{j \in hp(i)} \left\lceil \frac{R_i}{T_j} \right\rceil C_j, \tag{2.11}$$

where $hp(i)$ is the set of tasks with higher priorities than task i, and $\left\lceil \frac{R_i}{T_j} \right\rceil$ denotes the ceiling function.

Thanks to its simplicity in implementation and optimality in performance, the RM algorithm has gained wide support from most commercial real-time operating systems. Even the development of some real-time programming languages (e.g. real-time Java) and standards (e.g. real-time POSIX and UML) has been influenced by the theory related to this algorithm [WIT02]. While the original results were based on a set of rather restrictive assumptions, most of these have subsequently been removed or relaxed over the years [SHA04].

Dynamic-Priority Scheduling

In contrast to fixed-priority scheduling algorithms, a dynamic-priority scheduling algorithm assigns different priorities to individual jobs of each task. The priority of the task relative to other tasks changes as jobs are released and completed.

A commonly used algorithm belonging to this class is the Earliest-Deadline-First (EDF) algorithm, also presented in [LIU73]. It assigns priorities to individual jobs of the tasks according to their absolute deadlines, i.e. the highest priority is assigned to the task having the earliest absolute deadline. The necessary and sufficient schedulability condition for EDF is:

Theorem 2.3 A system that contains N independent, preemptive, periodic tasks with relative deadlines equal to periods (i.e. $D_i = T_i$) is schedulable according to the EDF algorithm if and only if the CPU utilization

$$U = \sum_{i=1}^{N} \frac{C_i}{T_i} \leqslant 1. \tag{2.12}$$

It has been showed that EDF is optimal among all preemptive scheduling algorithms [DER74]. This means that if there exists a feasible schedule for a task set, then it must be schedulable with EDF. More complex analysis on EDF schedulability also exists in which many of the assumptions above have been loosened.

There are many advantages with EDF scheduling [WIT02]. For example, an obvious advantage is that the processor can be fully utilized and still all deadlines can be met. Another advantage comes from the fact that it is often more intuitive to assign deadlines to tasks than to assign priorities. Assigning priorities requires global knowledge about the priorities of all the other tasks, whereas assigning deadlines only requires local knowledge.

Despite its theoretical advantages, EDF has mainly been used in research. There is a lack of commercial products supporting it, because EDF also has a number of potential drawbacks compared to FP scheduling [BUR01]. Some of them are: 1) the implementation of EDF is more complex; 2) it incurs a larger runtime overhead and requires more storage; 3) during overloads, all tasks tend to miss their deadlines.

2.2.2 Real-Time Communication

In real-time communication systems, the network becomes the resource that is shared by multiple communication entities. Therefore, the management of the communication resources is of special interest. *Messages* are exchanged among applications on different nodes. Typically, they are fragmented into segments for their transmission over the communication network. Each segment is handled by the network as a basic transmission unit usually called a *packet*, and the transmission of a packet is non-preemptive. In principle, the transmission of messages over networks is similar to the execution of tasks on processors. Like tasks, messages can be *periodic*, *aperiodic*, or *sporadic*. With regard to timing parameters, a message also has a (*worst-case*) *transmission time* and a *relative deadline*. The transmission of a message can be simply viewed as a task, with the transmission time being the execution time.

Typical communication networks are designed based on layered protocol stacks such as the seven-layer OSI (Open System Interconnection) reference model [TAN02]. It is the responsibility of the entire protocol stack to guarantee real-time communication. Problems like resource contention can generally be tackled at different levels in the protocol stack and different combinations of strategies can be developed. From a resource scheduling perspective, however, the network is simply a shared resource for which different messages have to compete. Further, the crucial role of coordinating the access from active nodes falls onto the MAC protocols [BEN06, PAL02].

In the communication literature, a large number of MAC protocols have been developed. Depending on how the nodes are allowed to access the medium, two categories of MAC protocols can be identified [KUM06]: controlled access and random access. In the first category the time and order of each node accessing the medium are controlled by specific algorithms. Since no collision will occur, this type of MAC protocols is also called *contention-free* MAC protocols. A typical example is TDMA (Time Division Multiplex Access), which is somewhat comparable to the offline scheduling of processors [PAL02]. Many fieldbuses such as Profibus, P-NET, IEC 1158, WorldFIP, Interbus, FF, etc. also belong to this category. On the contrary, with the second category of MAC protocols, each node can request medium access at arbitrary time, in a random manner, thus allowing collisions among different nodes. That is, these MAC protocols are *contention-based*. CSMA

(Carrier Sense Multiple Access)[1] and variants are typical examples of contention-based MAC protocols.

TDMA

The principle of TDMA[2] is quite simple. In TDMA, there is a base station that coordinates the nodes attached to the network. The time on the network is divided into *time slots*, which are generally of fixed size. Each node is allocated at pre-runtime a certain number of slots in which it can send and receive messages. In this way, multiple nodes are allowed to access a single shared network without collisions.

TDMA is used in the digital 2G cellular systems, the DECT (Digital Enhanced Cordless Telecommunications) standard for digital portable phones and satellite systems, among others. TDMA performs very well in achieving temporal determinism, which makes it well-suited for real-time applications. A variety of networks or communication protocols are built upon TDMA, for example, SERCOS, ARINC, and TTP. Because it is too strict and inflexible, however, TDMA is not suitable for data networking applications such as the Internet with unpredictable traffic [WIR07].

CSMA/CD

CSMA/CD stands for CSMA with collision detection, a set of rules for determining how network nodes respond in the event of a collision. Carrier sense refers to the fact that every node listens to the network before it attempts to transmit. If the network is busy, it waits until the network is idle. Multiple access means that more than one node can be sensing (listening and waiting to transmit) at a time. Collision detection means that when multiple nodes accidentally transmit at the same time, they can detect the collision.

CSMA/CD is internationally standardized in IEEE 802.3 and ISO 8802.3. Standard Ethernet[3] networks use CSMA/CD to resolve contention conflicts on the communication medium. On Ethernet, any node can try to send a packet at any time. If the network is idle, the sending node then begins to transmit. While transmitting, it also listens to detect if there is a collision. If a collision occurs, every participating node stops transmitting and waits for a random time interval to retry its transmission. This random time is determined by the standard *binary exponential backoff* (BEB) algorithm: the retransmission time is randomly chosen between 0 and $(2^i - 1)$ slot times, where i denotes the i-th collision event detected by the node and one slot time is the minimum time needed for a round-trip transmission. After 10 collisions have been detected, the interval is fixed at the maximum of 1023 slots. The relevant packet will be discarded after 16 collisions

---

1 http://en.wikipedia.org/wiki/Carrier_sense_multiple_access

2 http://en.wikipedia.org/wiki/Time_division_multiple_access

3 http://en.wikipedia.org/wiki/Ethernet

[LIA01, TAN02].

The main advantage of CSMA/CD is that it is simple to implement, which helps to make Ethernet a *de facto* standard for Local Area Networks (LANs). However, CSMA/CD is a non-deterministic protocol and does not support any message prioritization. Collisions become a major problem at high network workloads because time delay may become unbounded [WHE93].

CSMA/CA

CSMA/CA (CSMA with Collision Avoidance) is another modification of pure CSMA, where collision avoidance rather than collision detection is used to improve the performance of CSMA. It is designed to reduce the probability of collisions between multiple nodes accessing a medium, at the point where collisions would most likely occur. This is the underlying meaning of collision avoidance. Unlike CSMA/CD that deals with a collision after it has been detected, CSMA/CA acts to avoid collisions before they happen. This is the main difference between these two protocols.

CSMA/CA is mainly used where CSMA/CD cannot be implemented due to e.g. the nature of the communication medium. A typical example is the IEEE 802.11 protocol series [4], which specifies two medium access coordination functions, the mandatory Distributed Coordination Function (DCF) that is based on CSMA/CA and the optional Point Coordination Function (PCF). Unlike wired nodes, wireless nodes cannot detect collisions because they are half-duplex, i.e. they cannot send and receive signals at the same time. That is why is not applicable to wireless communications.

In IEEE 802.11 [IEEE99], each node senses the medium before starting a transmission. Carrier sensing is performed through both physical and virtual mechanisms. Time slots are divided into multiple frames and the time interval between frames is called the Inter-Frame Space (IFS). Four different IFSs are defined to provide priority levels for access to the wireless media, including Short IFS (SIFS), PCF IFS (PIFS), DCF IFS (DIFS) and Extended IFS (EIFS), ordered from the shortest to the longest.

If the medium is idle for at least a DIFS, the packet is transmitted immediately. If the medium is sensed busy, the node waits for the end of the current transmission and then starts the contention, also called backoff process. The node selects a random backoff time given by: BackoffTime = Random( ) × SlotTime, where Random() is a pseudo random integer drawn from a uniform distribution over the interval [0, CW], where the contention window (CW) is an integer within the range of values of the physical-layer characteristics CW_{min} and CW_{max}, i.e. $CW_{min} \leqslant CW \leqslant CW_{max}$, and SlotTime is the value of the correspondingly named physical-layer characteristic. The CW is incremented exponentially by the increasing number of attempts to retransmit the packet.

---

4 http://standards.ieee.org/getieee802/802.11.html

During the backoff process, the backoff timer is decremented in terms of slot time as long as the medium is idle. When the medium is busy, the timer is frozen. When its backoff timer expires, if the network is still idle, the data packet is sent out. The node having the shortest contention delay wins and transmits its packet. The other nodes just wait for the next contention (at the end of the transmission of this packet). If another collision occurs, a new backoff time is chosen and the backoff procedure starts over again until sometime the limit is exceeded. Because the contention is based on a random function and done for every packet, each node is thus given an equal chance to access the communication medium.

Upon receipt of an intact packet, the receiving node waits for a SIFS interval and transmits a positive acknowledgment frame (ACK) back to the source node, indicating transmission success. If the source node does not receive an ACK, it assumes that the packet was collided, and attempts a retransmission and enters again the backoff procedure. To reduce the possibility of collisions, the CW is doubled up to a limit $CW_{\max}$ after each retransmission attempt. It is reset to $CW_{\min}$ after each successful transmission. In the presence of an error, a packet has to be retransmitted by the source node.

CSMA/CA delivers a best effort service, thereby providing no bandwidth and delay guarantees. Like CSMA/CD, it is a non-deterministic protocol, but there are also many advantages. For instance, CSMA/CA is well suited for network protocols such as TCP/IP, adapts rather well with the variable conditions of traffic, and is quite robust against interferences [WIR07].

CSMA/BA

The CSMA/BA (CSMA with Bitwise Arbitration) protocol is typically used by CAN (Controller Area Network)[5] [BOS91] and DeviceNet to provide collision resolution. This protocol is sometimes referred to as a special kind of CSMA/CD or CSMA/CA, or called CSMA/AMP (CSMA with Arbitration on Message Priority) [LIA01].

Like CSMA/CD and CSMA/CA, each node on a network using CSMA/BA must wait for a prescribed period of inactivity before starting to send a message. Once this period of inactivity elapses, every node on the network can transmit a message. Bitwise Arbitration means that collisions are resolved through a bit-wise arbitration, based on a pre-programmed priority of each message in the *identifier* field. For CAN the identifier is a unique sequence of 11 (standard CAN frame format, version 2.0A) or 29 bits (extended CAN frame format, version 2.0B). It assigns a priority to the message and enables the receivers to filter messages. Since CAN is a broadcast network, messages do not explicitly contain addresses of either the source or the destination.

The arbitration mechanism of CAN works as follows [BEN06, TOV99]. A smaller binary message identifier number always indicates a higher priority. A CAN

[5] http://en.wikipedia.org/wiki/Controller_Area_Network

message associated with the highest priority will win the arbitration. This is achieved by CAN transmitting data through a binary model of *dominant* bits and *recessive* bits where dominant is a logical "0" and recessive is a logical "1". A dominant bit always overwrites a recessive bit on a CAN bus. If multiple nodes are transmitting simultaneously and one node transmits a "0", then all nodes monitoring the network will read a "0". Only if all nodes transmit a "1" will all nodes read a "1". The CAN behaves like a logical AND gate. During the transmission of the identifier field, if a node transmits a "1" and reads a "0", it means that there was a collision with at least one higher-priority message, and consequently this node aborts the message transmission. The highest-priority message being transmitted will proceed without perceiving any collision, and thus will be successfully transmitted.

The arbitration process of CAN is non-destructive [TOV99], in the sense that the node winning arbitration just continues on with the message transmission, without the message being destroyed or corrupted by any other nodes. The determinism of CAN makes it particularly attractive for use in Real-Time Control Systems.

The CAN in its current version 2.0 is an international standardized scalable serial bus communication system, originally developed by BOSCH in the mid-1980s for automotive applications. Since its conception, the CAN protocol has gained widespread popularity in industrial automation, medical control, automotive applications, etc.

The major disadvantage of CAN compared with the other networks is the slow data rate [LIA01]. This is mainly caused by the bit-wise arbitration mechanism, which poses strict limitations on the physical characteristics of the network including length and transmission data rate. For instance, transmission data rates up to 1 Mbps are possible if the network length is less than 40 m. For longer network distances, the data rate decreases, e.g. 125 kbps at 500 m.

2.3 Control Loop Timing

From a real-time scheduling point of view, a control task can generally be modelled as a periodic task, with a period equal to the sampling period of the relevant control loop. By default, it is assumed that the relative deadline equals the period for a control task. By a *control task*, we may refer to a software program executing a control algorithm in an embedded control system (from Chapters 3 to 6), or a message to be transmitted over the network in a (wired/wireless) networked control system (Chapters 7 and 8), or even both of them wherever appropriate. Accordingly, the execution time of a control task could be the computational delay of the control algorithm, or the transmission time of the packet, or even the delay of the whole loop.

The operations within a control loop have been observed in Section 2.1.1. The timing problems of a control loop contain the delay and its jitter, the sampling period and its jitter, and the data loss. These problems have attracted extensive attention in the control community [SAN00, TIP03, WIT95]. In this section we examine the temporal attributes of control loops from a real-time systems perspective. Also identified are some related aspects of the system environments considered in this book.

2.3.1 Delay and Its Jitter

Delay is one of the most fundamental timing parameters in real-time systems, referred to as response time. In a control loop, *delay* is the time interval between a sampling instant and the corresponding actuating instant, also known as *control delay* or *input-output latency*. The delay can be schematically described by Fig. 2.3, also see Fig. 2.2.

The delay τ_k comes mainly from the processing delay of the A/D converter, the communication delay between the sensor and the controller, the computational delay of the control algorithm, the communication delay between the controller and the actuator, and the processing delay of the D/A converter. Delay decreases the phase margin of a control system, which in turn deteriorates the system stability. Therefore, the delay should be as small as possible in favour of high QoC.

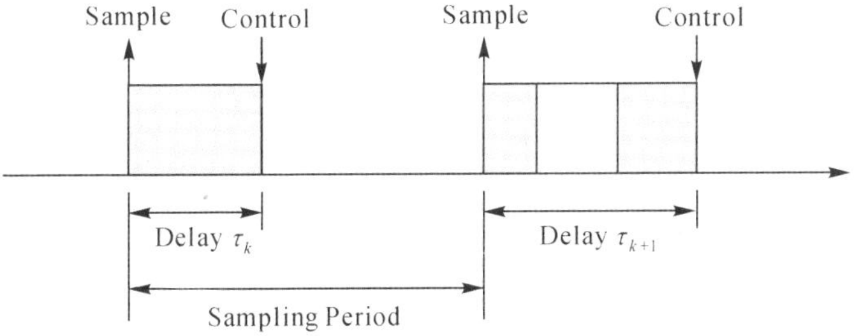

Fig. 2.3 Schematic diagram of delay and its jitter

If the delay is constant, it is quite easy to compensate for it in controller design [ARZ99, AST97]. In practice, however, this is not always the case. The delay may vary from one sample to another, giving rise to *delay jitter*. The delay jitter is caused by e.g. the contention for computing and/or communication resources among multiple tasks, the data dependency of control algorithms and temporal non-determinism of the implementation platforms. In addition, scheduling multiple tasks on a shared resource would potentially induce jitter in the delay.

In accordance with the definition of IEEE [IEEE92], jitter is *time-related, abrupt, spurious variations in the duration of any specified related interval*. It is a commonplace phenomenon in real-time systems. A straightforward understanding is that delay jitter is the variations in delay. As illustrated in Fig.2.3, delay jitter $J_\tau(k) = \tau_{k+1} - \tau_k$. Sometimes it is defined as the difference between the maximum and the minimum delays in all task instances, i.e. $J_\tau = \max\{\tau_i\} - \min\{\tau_i\}$.

In control systems, delay jitter degrades control performance. The degradation is closely related to the dynamics of the physical process, the control algorithm used, the type and magnitude of the jitter, etc. Generally, delay jitter should be as small as possible in order not to deteriorate QoC.

There are four common approaches to deal with variable delays [ARZ99]. The most straightforward approach is to implement the system in a way that the delay is minimized and then ignore it in the controller design. In resource constrained, multitasking environments, however, the efficiency of this approach is limited.

A second approach is to attempt to maintain a constant delay by means of, e.g. proper system architecture design and network design, and then account for this delay in the controller design. For instance, it is possible to introduce a queuing mechanism into the control loop to ensure that the delay is always equal to a specific value [TIA06]. When the control algorithm is designed in discrete-time domain, it is simple to take into account a constant delay that is a fraction of the sampling period in most controller design methods. For this purpose, a discrete-time description of the controlled system is obtained by sampling the continuous-time model containing a control delay [AST97]:

$$\begin{aligned} \frac{\mathrm{d}x}{\mathrm{d}t} &= Ax(t) + Bu(t-\tau) \\ y(t) &= Cx(t) + Du(t) \end{aligned} . \tag{2.13}$$

The sampled system is described by:

$$\begin{aligned} x(kh+h) &= \Phi x(kh) + \Gamma_0 u(kh) + \Gamma_1 u(kh-h) , \\ y(kh) &= Cx(kh) + Du(kh) , \end{aligned} \tag{2.14}$$

where

$$\begin{aligned} \Phi &= \mathrm{e}^{Ah} \\ \Gamma_0 &= \int_0^{h-\tau} \mathrm{e}^{As} B \mathrm{d}s \\ \Gamma_1 &= \int_{h-\tau}^{h} \mathrm{e}^{As} B \mathrm{d}s \end{aligned} . \tag{2.15}$$

A third approach is to treat the variable delay as an uncertain parameter of the system and design a controller that is robust to this uncertainty, e.g. [LI05, LI06]. This approach often depends on *a priori* knowledge about the distribution of the delay. For complex physical processes, it could be fairly more difficult.

The last approach relies on the controller to actively compensate for (all or parts of) the delay in each sample [NIL98]. For example, a simple scheme that compensates for the delay in the previous sample is presented in [LIN02]. To enable this approach, however, the delay in each sample has to be gathered at runtime, which induces extra computation and communication overheads.

From a real-time scheduling point of view, a delay larger than the sampling period implies that the task misses its deadline (which equals its period), see Figs. 2.2 and 2.3. The delay will be no more than the sampling period, though possibly time-varying, as long as the system schedulability is respected. Since one of the objectives of resource management is to maintain system schedulability, the (time-varying) delays smaller than the sampling period will be of major concern for systems considered in this book. With such a type of delays, the stability of the control system could be examined using established control theory, e.g. [KAO04, KIM03].

Since controller design is not the focus of this book, compensation methods for delay and delay jitter will not be used. A natural advantage of this is that the real effects of delay and its jitter on QoC will be substantially reflected in simulation experiments. Actually, compensating for a delay smaller than the sampling period is not difficult. Many related theories and methods could be found in the control community.

2.3.2 Sampling Period and Its Jitter

The sampling period is undoubtedly an important design parameter for sampled-data control systems. In control systems engineering, the sampling period is often determined based on the dynamics of the physical process to be controlled, desired closed loop performance, the control algorithm used, etc.

For a discrete-time design, a commonly used rule-of-thumb for choosing the sampling period is that the number of sampling instances per rise time should be between 4 and 10 [AST97], i.e.

$$4 \leqslant \frac{T_{rise}}{h} \leqslant 10, \tag{2.16}$$

where T_{rise} is the rise time of the resulting closed loop system.

When the discretization based design method is used, the basic rule for choosing the sampling period is the shorter the better given that the required support from the underlying platform is available. Some rules-of-thumb can also be used. An example rule-of-thumb for second-order systems is [WIT02]

$$0.05 \leqslant h\omega_c \leqslant 0.14, \tag{2.17}$$

where ω_c is the cross-over frequency (rad/s) of the continuous-time system.

It can be seen from Eqs.(2.16) and (2.17) that, no matter which design method is used, control systems can always operate over quite a large range of sampling periods while delivering good performance. This feature of sampled-data control systems makes it possible to apply feedback scheduling. We will come back later (in Section 2.5) on this topic.

Generally the larger the sampling period is, the worse the control performance becomes. To meet the application requirements on system stability or desired QoC, it is mandatory that the sampling period should not exceed its maximum allowable upper bound h_{max}. Methods for obtaining the maximum allowable sampling period that guarantees the stability of a control system are available in the control community, e.g. [MON04, ZHA01].

When the control algorithm is implemented as a real-time task, the sampling period naturally becomes the period of the control task. Therefore, the sampling period influences not only the QoC but also the resource demands of the task.

Ideally, the output of the physical process is sampled equidistantly in time, i.e. at time instants $t = 0, h, 2h, \cdots, kh$. However, the sampling instants in real systems may be different, implying that the sampling is not equidistant. This is particularly true when resource contention, clock inaccuracy, and workload variations are present. A natural result is that the actual sampling period varies from sample to sample. These variations of sampling period are known as *sampling period jitter*, i.e. $J_h(k) = h_{k+1} - h_k$(see Fig.2.4). Similar to delay jitter, the sampling period jitter is sometimes defined as the difference between the maximum and the minimum sampling periods, i.e. $J_h = \max\{h_i\} - \min\{h_i\}$.

It has been found that relatively small sampling period jitter need not be compensated for, since its impact on control performance is not significant [CER00, SAN04]. Generally speaking, systems that use slow sampling and those with small phase margins are more sensitive to sampling period variations [AST97].

It is possible to distinguish between unintentional and intentional changes in the sampling period. The sampling period jitter demonstrated in Fig.2.4 belongs to the first class. In this book we do not use any methods to cope with unintentional sampling period jitters. Nevertheless, the sampling period will be adjusted intentionally within the framework of feedback scheduling. It is then necessary to compensate for these intentional sampling period variations. This will be detailed in Section 2.5.

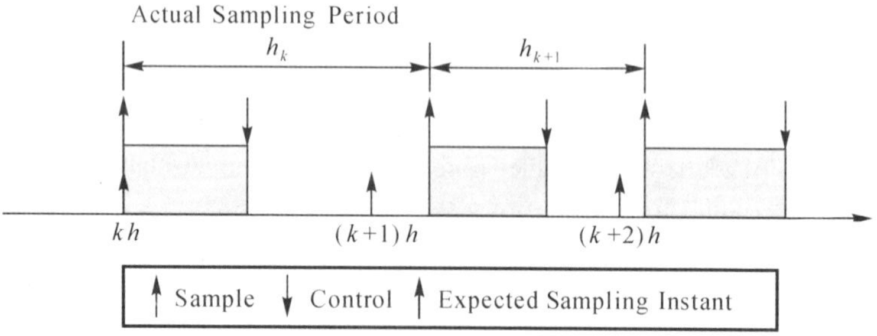

Fig. 2.4 Schematic diagram of sampling period jitter

2.3.3 Data Loss

When the data flows (from the sensor to the actuator) within a real-time control loop given in Fig.2.1, it is possible that it gets lost, as observed in the third sampling interval of Fig.2.2. As a result, the actuator will never receive a control command associated with the relevant sample. This phenomenon is called *data loss*, which directly leads to *vacant sampling*. In the networked control literature, data loss is also referred to as *packet loss* or *packet dropout* since no other types of data losses are considered. A data loss occurs when the measurement or the control command is lost during the process of transmission on the network, or the control task fails in producing the control input signal.

In the presence of data losses, the control system is effectively run in open loop. Therefore, the QoC of the system will certainly be degraded. The number of lost data should be kept as small as possible, ideally to be zero. It can be argued that most practical control systems are able to tolerate data losses to a certain degree. Of course, the data loss ratio should be within a certain limit in order to maintain system stability. Theoretical methods for computing the upper bound of data loss ratio can be found in e.g. [AZI03, LIN03a, ZHA01].

From the real-time scheduling perspective, the execution time of a task becomes infinite in the event of data loss. Such an execution time implies that the deadline of the task is certainly missed. From this insight, data loss can be viewed as a special kind of deadline miss, though it may not be a direct consequence of system unschedulability.

The most straightforward way to deal with data losses is to minimize the probability of data being lost. One possibility is to reduce the number of packet dropouts on the network through retransmitting the data. However, when the bandwidth is limited, too many retransmissions may cause unexpected congestions that drive the problem from bad to worse. With the penalty of increased delay, this may adversely deteriorate the control performance.

Another way is to compensate for the data loss after it occurs. Possible methods are for example adding error correction to the transmitted measurement and control command packets, modifying the state estimation in the controller, modifying the control algorithm explicitly, and jointly designing the estimator and the controller [ARZ06b, LIN03b, SCH06]. A simple method for dealing with data loss will be presented in Chapter 8.

2.4 Motivating Examples

In this section we examine the effects of resource scheduling on the control performance through simulating a concrete case. To simplify the description, only the scheduling of the CPU in a multitasking embedded control system is considered here.

Assume that all control loops in the system are independent of each other. The controlled processes are the DC servo system modelled as:

$$G(s) = \frac{1000}{s^2 + s}. \tag{2.18}$$

The controllers in all control loops use the PD (Proportional-Derivative) algorithm with the same form and parameters as [OHL06]. The timing parameters, i.e. sampling period h and execution time c, of each control task are given in Table 2.1.

Table 2.1 Timing parameters for simulations

	h (ms)	c (ms)
Loop 1	6	2
Loop 2	5	2
Loop 3	4	2

Example 2.1 The RM scheduling policy is used in the system, where there are only two loops, i.e. Loops 1 and 2. Fig.2.5 gives the corresponding step responses of the two control loops as well as the schedule of CPU.

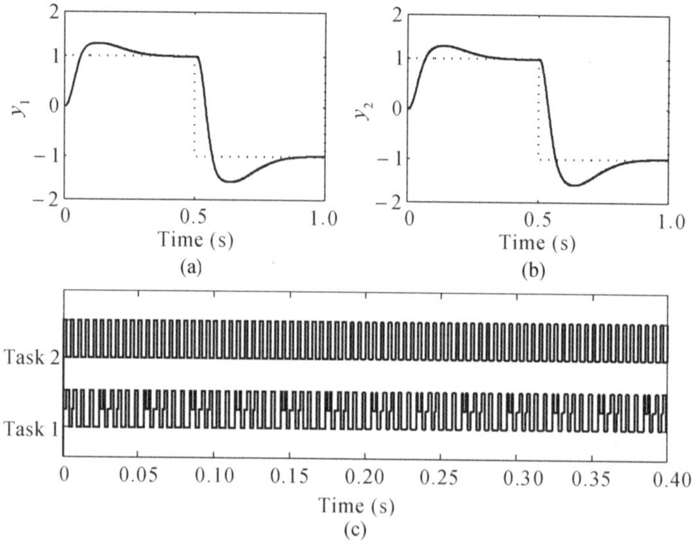

Fig. 2.5 Two control loops under RM scheduling
(a) Loop 1; (b) Loop 2; (c) CPU schedule

As Fig.2.5 shows, when there are only two control loops competing for the CPU resource, both loops perform very well. The corresponding CPU utilization is $2/6 + 2/5 = 0.73 < 2(2^{1/2} - 1) = 0.83$. It is obvious that the system is schedulable, see Theorem 2.1. This can also be seen from the CPU schedule. Moreover, since $h_1 > h_2$, which implies that the priority of Task 2 is higher than that of Task 1, the execution of Task 1 may possibly be preempted by Task 2 during runtime.

Example 2.2 The system uses the RM scheduling policy, while three control loops are running in parallel. The corresponding system responses and CPU schedule are shown in Fig.2.6.

The instability of the system is clearly shown in Fig.2.6. Despite the good performance of Loops 2 and 3, Loop 1 becomes unstable. According to the RM policy, Task 1 holds the lowest priority because of its largest sampling period. The CPU utilization is $2/6 + 2/5 + 2/4 = 1.23 > 1$, which means that the system is overloaded and consequently unscheduable. As can be seen from the CPU schedule,

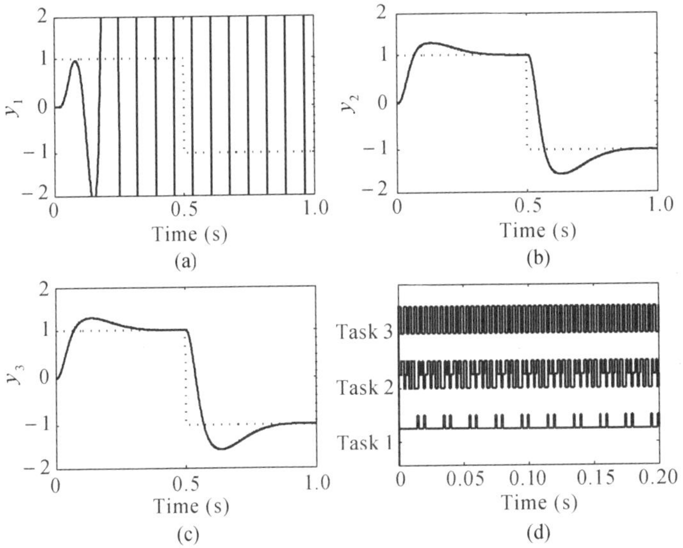

Fig. 2.6 Three control loops under RM scheduling
(a) Loop 1; (b) Loop 2; (c) Loop 3; (d) CPU Schedule

control Task 1 is continuously preempted by the other two high-priority tasks most of the time. This causes a high deadline miss ratio in Loop 1, thus leading to system instability. In contrast, the execution of Task 3 will not be affected by Tasks 1 and 2 thanks to its highest priority, whereas Task 2 is still possible to be preempted by Task 3.

Example 2.3 The EDF scheduling policy is used to schedule the three control tasks running concurrently. The corresponding system responses and CPU schedule are depicted in Fig.2.7.

As we can see from Fig.2.7, the system is stable. The requested CPU utilization of three tasks is still 1.23 > 1, which indicates system unschedulability (see Theorem 2.3), but the performance of the system remains satisfactory. The main reasons behind can be explained as follows:

(1) In contrast to Example 2.2 where almost all deadline misses are experienced by Task 1, under EDF all three control tasks (rather than only one task) are likely to miss their deadlines, see the CPU schedule in Fig.2.7. This keeps the deadline miss ratio of Loop 1 within a tolerable range, while the deadline miss ratios of the other two loops will also not be too high.

(2) Because the upper bound of schedulable utilization for EDF (i.e. 100%) is larger than that of RM, the overall deadline miss ratio of the system will decrease relative to Example 2.2.

The CPU schedule also shows that all three control tasks have gained sufficient CPU time to execute, thereby resulting in satisfactory performance. In [CER02] Cervin, *et al.* have discussed the performance of EDF under overload conditions. Therefore, detailed discussion on this issue is intentionally skipped here.

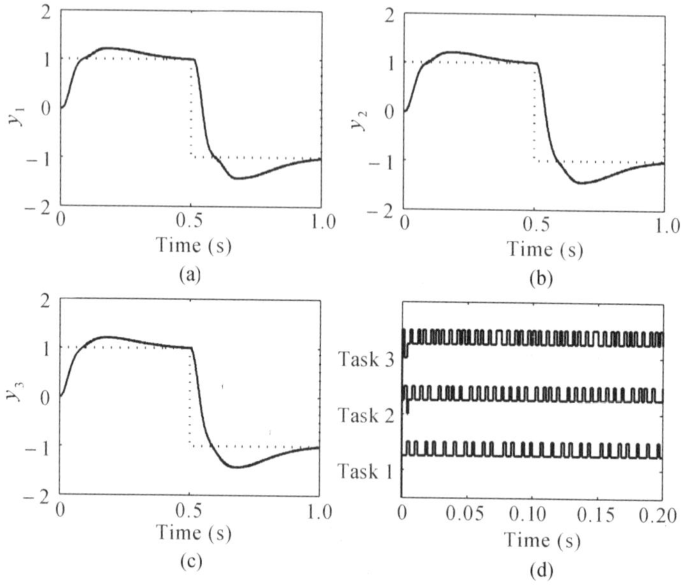

Fig. 2.7 Three control loops under EDF scheduling
(a) Loop 1; (b) Loop 2; (c) Loop 3; (d) CPU Schedule

To summarize the above three examples, we note that:

(1) The traditional system design method can achieve good performance under resource-sufficient conditions, see Fig. 2.5. When the resource becomes insufficient, however, the system performance may be significantly degraded, even destabilized in extreme cases, as shown in Fig.2.6.

(2) Traditional open-loop scheduling algorithms cannot distribute resources among control systems in an optimal way. Although RM and EDF are optimal in the sense of real-time scheduling, they do not take into account the actual requirements of the target applications, and consequently are unable to optimize the overall QoC of the system.

(3) Different resource scheduling policies may possibly deliver rather different overall system performance, as can be seen from the comparison of Figs.2.6 and 2.7. This indicates that it is possible to optimize/improve the overall system performance by means of resource management.

Therefore, it is necessary to develop a new kind of resource management schemes for Real-Time Control Systems, which may be closed loop and associated with system performance such that the QoC could be managed flexibly.

Example 2. 4 Revisit the system in Example 2.2, and change the sampling periods to $h_1 = 9$ ms, $h_2 = 8$ ms, and $h_3 = 7$ ms. The corresponding system responses and CPU schedule are given in Fig.2.8.

Compared with Example 2.2, the sampling periods of three control loops are enlarged in Example 2.4. According to sampled-data control theory, an increase of sampling period may degrade control performance. However, the system clearly

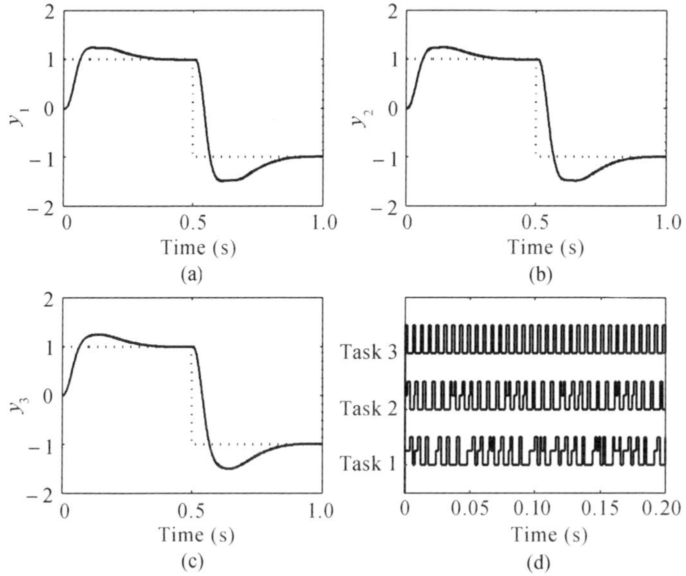

Fig. 2.8 Three control loops with new sampling periods
(a) Loop 1; (b) Loop 2; (c) Loop 3; (d) CPU Schedule

achieves better overall performance in this case. All control loops remain stable with good performance all the time. Fig.2.9 depicts the total control costs (i.e. the sum of each loop's Integral of Absolute Error) in both Example 2.2 and Example 2.4. It is apparent that the overall system performance of Example 2.4 is much better than that of Example 2.2.

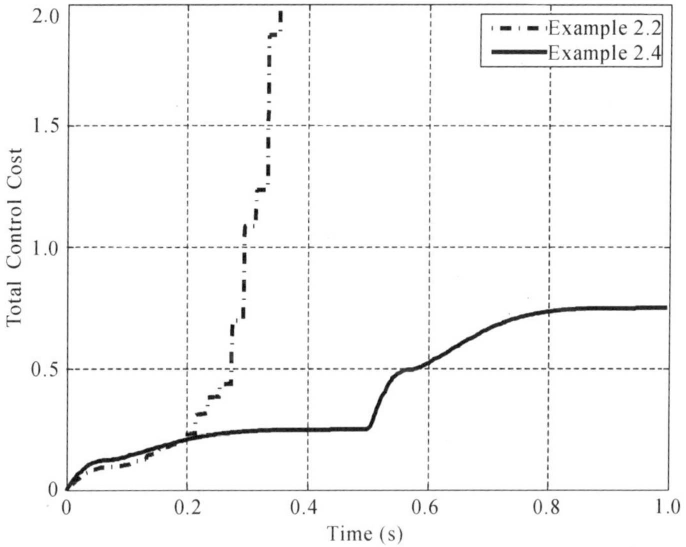

Fig. 2.9 Total control costs with different sampling periods

In Example 2.2 we have pointed out that the main reason for the system instability is the unschedulability of the set of control tasks. On the contrary, the (requested) CPU utilization in Example 2.4 becomes $2/9 + 2/8 + 2/7 = 0.73 < 3(2^{1/3} - 1) = 0.78$, which means that the system is schedulable, see Theorem 2.1. Although increasing sampling periods will degrade the control performance to some degree, the control tasks' demands on resources are also reduced at the same time, thus preserving the system schedulability. Deadline misses are consequently avoided. That is why the resulting system performance is improved in Example 2.4.

A preliminary conclusion that can be made based on the above results is that in resource insufficient environments it is possible to improve the overall control performance by properly adapting some timing parameters such as sampling periods of the control tasks. This constitutes the basic idea that motivates feedback scheduling.

2.5 Feedback Scheduling

Feedback scheduling offers a promising approach to flexible resource management in Real-Time Control Systems [ARZ06a, XIA06b]. Exploiting the methodology of feedback, feedback schedulers regulate online timing parameters of control tasks in response to the continuous feedback of information about the actual resource usage. The integration of feedback control and real-time scheduling is explored to deal with uncertainties in available resource and workload, thus providing guarantees of QoC for the system.

2.5.1 Basic Framework

Feedback scheduling can broadly speaking be described as feedback control based dynamic resource scheduling. That is, feedback schedulers usually adapt the execution of a set of real-time tasks according to feedback information about actual usage of the resource. In principle, the resource referred to here could be any kind of shared resources, typical of which are CPU time, energy and network bandwidth. Accordingly, the execution of tasks could be the run of control software programs in embedded control systems, or the transmission of data packets in wired networked control systems or wireless control systems. The most distinct characteristic of feedback scheduling is that it introduces feedback into dynamic resource management. In a sense, feedback scheduling can be considered as a product of the interplay of feedback control and real-time scheduling.

In this book we study feedback scheduling theories and methods intended for Real-Time Control Systems, using feedback scheduling as the underlying tool for the codesign of control and scheduling. As mentioned in Chapter 1, on the one hand, we dynamically manage the limited resources in Real-Time Control Systems by employing the feedback scheduling technology, such that the overall system performance is optimized. On the other hand, we exploit feedback control principles and techniques when designing feedback schedulers. There is no exaggeration that feedback scheduling has become one of the most important enabling technologies for the integration of computing, communication and control.

Fig.2.10 gives a typical framework of feedback scheduling [XIA06a, XIA06b]. In the system multiple control loops share the same computing and/or communication resources, i.e. the processor and/or the network. Besides these control loops, an outer feedback loop realizing feedback scheduling is added. The basic role of the feedback scheduler is dynamically adjusting the related scheduling parameters (i.e. timing parameters) of control loops/tasks to achieve a desired resource utilization or deadline miss ratio. The feedback scheduling loop resembles a classical feedback control loop. Naturally, control theory and technology may be applied to the design of feedback schedulers. In the literature, a feedback scheduling system designed using a control theoretic approach is sometimes referred to as feedback control (real-time) scheduling, which is regarded as a special class of feedback scheduling in this book.

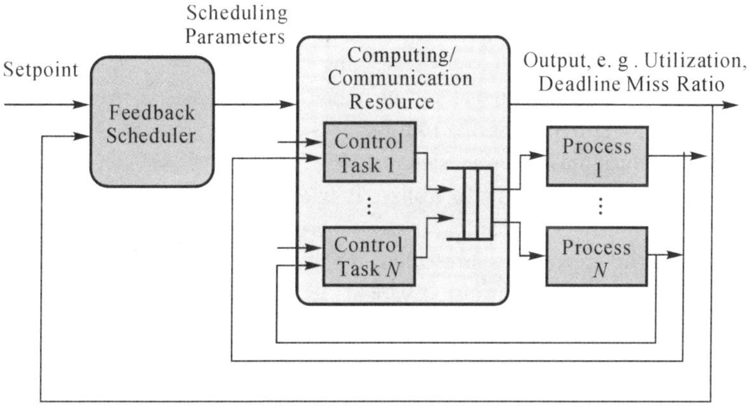

Fig. 2. 10 Basic framework of feedback scheduling

Notice that the feedback scheduling schemes to be presented in this book are not actually real-time scheduling algorithms/policies. None of the feedback schedulers is a *scheduler* in real-time scheduling terms (see Section 2.2). They are instead resource managers that run on top of existing real-time scheduling policies such as RM and EDF.

2.5.2 Design Issues

A prerequisite for the design of a feedback scheduler that can be treated as a controller is to properly select the control-related variables, among which the most important ones are the *controlled variable* and the *manipulated variable*.

For real-time tasks sharing computing or communication resources, the most common performance measures are resource utilization and deadline miss ratio. The controlled variable of the feedback scheduling loop may accordingly be chosen to be either resource utilization or deadline miss ratio. When the utilization is selected, the setpoint for the utilization should generally be maximized as much as possible to make full use of available resources, given that the system schedulability is preserved. On the contrary, when the deadline miss ratio is selected, small setpoints are preferable in order for reduction of the negative effects of deadline misses on control performance, given that the effectiveness of feedback scheduling is guaranteed. Notice that the range of the possible values of resource utilization and deadline miss ratio is [0, 100%]. This indicates that in feedback scheduling systems the resource utilization and the deadline miss ratio saturate at 100% and 0, respectively, which should be taken into consideration at design time.

According to the real-time task model commonly used in real-time scheduling theory, see Section 2.2, options for the manipulated variable are the timing parameters of tasks including period, execution time, (relative and absolute) deadlines, etc. In feedback scheduling of control tasks, the sampling period (i.e. task period) is the most common choice for the manipulated variable. There are some reasons as follows:

(1) The sampling period affects simultaneously the resource demand of the control task and the resulting control performance of the loop;

(2) Almost all real-world control systems can perform well with a variety of sampling periods within a certain range, which makes it possible to adapt the sampling period during runtime;

(3) As a design parameter in sampled-data control systems, the adaptation of sampling period is easy to realize. It is also convenient for the controller to compensate for the changes in sampling period.

In most of the approaches presented in this book, the sampling period will be chosen as the (primary) manipulated variable in the feedback scheduling loop.

When anytime control algorithms that comply with the imprecise computation model are used in control loops, the task execution time then becomes a better choice for the manipulated variable [ARZ03]. In Chapters 5 and 6, we will explore a novel method for adjusting task execution time, where the execution times of all tasks are indirectly adjusted at the same time by scaling the CPU speed based on the Dynamic Voltage Scaling technique.

The objective of using task deadline as the only manipulated variable usually relates to minimization of delay and jitter, e.g. [ZHO05], which is out of the scope of this book. It should be noted that the adjustment of a control task's period implies the simultaneous adjustment of its deadlines.

As pointed out previously, in order not to jeopardize the system stability, the constraint on the maximum allowable sampling period must be taken into account when adapting the sampling period. That is, it must be held that $h \leqslant h_{max}$. Theoretically, the maximum allowable sampling period for each control loop could be determined using related control theory concerning stability conditions, e.g. [MON04, ZHA01].

In this book we instead use a simple and practical method, i.e. the simulation-based approach to choose a maximum allowable sampling period for each control loop. Theoretical results in existence will be referred to only when necessary. Some reasons for the use of the simulation-based approach are outlined below:

(1) Stability analysis of control systems with variable sampling periods is quite complicated; available theoretical results, usually based on some idealized assumptions (e. g. no delay jitter exists), are only applicable to specific systems.

(2) Control system stability analysis is not the focus of our interest.

(3) In most situations the simulation-based approach can yield results that are more practical and hence more useful.

Since the feedback scheduler adapts the sampling periods of control loops during runtime, it is an obligation of the controllers to compensate for these changes in sampling periods. From an implementation viewpoint, it is possible for simple control algorithms such as PID and state feedback control that the controller parameters are updated directly in response to the changes in sampling periods. For complex control algorithms demanding a large amount of computations, it should be better to handle this in another way: first design offline different controllers for different sampling periods, store related controller parameters, then, during runtime, use the look-up table approach to select the most proper controller parameters for the current sampling period. The overheads associated with different implementation methods have been discussed by Marti [MAR02]. We choose to compensate for the changes in sampling periods induced by feedback scheduling through updating online the controller parameters, assuming that the extra overhead of this update is negligible.

Still another important issue is the overhead associated with feedback scheduling. It is intuitive that the feedback scheduler itself consumes resources during runtime. This feedback scheduling overhead should be minimized, particularly in resource-constrained environments, because the practical applicability of the feedback scheduler will be impaired if the overhead is excessively large. For this insight, we will always attempt to reduce the complexity of the feedback scheduler to be designed, given that the comparable performance can be achieved. In practice there is indeed a compromise that should be made between the performance and the overhead when applying feedback scheduling.

To avoid confusion of traditional control loops and the feedback scheduling loop, throughout this book we use the term *invocation interval* to represent the activation period of feedback schedulers, and *sampling period* to represent the operating period of control loops. In addition, the notation j is used in equations to denote the invocation instant of feedback schedulers, while k is used to denote the sampling instant of control loops.

2.6 Summary

This chapter has introduced key concepts and principles in both real-time control and real-time scheduling, providing the background knowledge necessary for understanding the methodology of feedback scheduling. Instead of a comprehensive overview of these fields, which is of course impossible, only selected topics have been covered that relate closely to the present work. The timing parameters of control loops have been described. The importance of the role that resource management plays in Real-Time Control Systems has been demonstrated via a set of motivating examples, thus illustrating experimentally the motivations for applying feedback scheduling. Key concepts in feedback scheduling have been explained. A basic framework has also been given along with discussions on related design issues.

This chapter provides a unified framework for our work. In the subsequent six chapters, we will study different aspects of dynamic resource management in Real-Time Control Systems, and develop a set of feedback scheduling schemes based on this unified framework.

References

[ARZ99] K.E. Årzén, B. Bernhardsson, J. Eker, A. Cervin, P. Persson, K. Nilsson, L. Sha. Integrated Control and Scheduling. *Research Report*, ISSN 0820 - 5316, Dept. Automatic Control, Lund Institute of Technology, 1999.

[ARZ03] K.E. Årzén, A. Cervin, D. Henriksson. Resource-Constrained Embedded Control Systems. *Possibilities and Research Issues*, *Proc. of Codesign of Embedded Real-Time Systems Workshop*, Porto, Portugal, 2003.

[ARZ06a] K.E. Årzén, A. Robertsson, D. Henriksson, M. Johansson, H. Hjalmarsson, K.H. Johansson. Conclusions of the ARTIST2 Roadmap on Control of Computing Systems. *ACM SIGBED Review*, Vol.3, No.3, pp.11 - 20, 2006.

[ARZ06b] K.E. Årzén, A. Bicchi, S. Hailes, K.H. Johansson, J. Lygeros. On the Design and Control of Wireless Networked Embedded Systems. *Proc. of IEEE Computer Aided Control Systems Design Symposium*, pp.440 - 445, 2006.

[AST97] K.J. Åström, B. Wittenmark. *Computer Controlled Systems: Theory and Design*, *3rd Edition.* Prentice Hall, 1997.

[AUD91] N.C. Audsley, A. Burns, M. Richardson, A. Wellings. Hard real-time scheduling: the deadline monotonic approach. *Proc. of IEEE Workshop on Real-Time Operating Systems and Software*, pp.133 - 137, 1991.

[AZI03] B. Azimi-Sadjadi. Stability of Networked Control Systems in the Presence of Packet Losses. *Proc. of IEEE Conf. on Decision and Control*, pp.676 - 681, 2003.

[BEN06] M.M.B. Gaid. Optimal Scheduling and Control for Distributed Real-Time Systems. Ph.D Thesis, COSI Laboratory, ESIEE, Paris, 2006.

[BOS91] BOSCH. CAN Specification (Version 2.0), 1991.

[BUR01] A. Burns, A. Wellings. *Real-Time Systems and Programming Languages, 3rd Edition*. Addison-Wesley, 2001.

[CER00] A. Cervin. Towards the Integration of Control and Real-Time Scheduling Design. Licentiate Thesis. Department of Automatic Control, Lund Institute of Technology. Lund, Sweden, 2000.

[CER02] A. Cervin, J. Eker, B. Bernhardsson, K. E. Årzén. Feedback-Feedforward Scheduling of Control Tasks. *Real-Time Systems*, Vol. 23, No.1, pp.25 – 53, 2002.

[DER74] M. L. Dertouzos. Control Robotics. The Procedural Control of Physical Processes. *Information Processing*, Vol. 74, 1974.

[HEL04] J.L. Hellerstein, Y. Diao, S. Parekh, D. Tilbury. Feedback Control of Computing Systems. Wiley-Interscience, 2004.

[IEEE92] IEEE Standard 100 – 1992. *The New IEEE Standard Dictionary of Electrical and Electronics Terms, 5th Edition.* The IEEE, Inc., 1992.

[IEEE99] International Standard ISO/IEC 8802 – 11, ANSI/IEEE Std 802.11, Part 11. *Wireless LAN Medium Access Control (MAC) and Physical Layer (PHY) Specifications.* The IEEE, Inc., 1999.

[JOS86] M. Joseph, P. Pandya. Finding Response Times in a Real-Time System. *BCS Computer Journal*, Vol. 29, No.5, pp.390 – 395, 1986.

[KAO04] C.Y. Kao, B. Lincoln. Simple Stability Criteria for Systems with Time-Varying Delays. *Automatica*, Vol. 40, No.8, 2004.

[KIM03] D.S. Kim, Y.S. Lee, W.H. Kwon, H.S. Park. Maximum Allowable Delay Bounds of Networked Control Systems. *Control Engineering Practice*, Vol.11, pp.1301 – 1313, 2003.

[KUM06] S. Kumar, V.S. Raghavan, J. Deng. Medium Access Control Protocols for Ad Hoc Wireless Networks: A Survey. *Ad Hoc Networks*, Vol. 4, pp.326 – 358, 2006.

[LEH89] J.P. Lehoczky, L. Sha, D.Y. Ding. The Rate-monotonic Scheduling Algorithm: Exact Characterization and Average Case Behaviour. *Proc. of IEEE Real-Time Systems Symposium*, pp.166 – 171, 1989.

[LI05] S.B. Li. Study on Robust Control Algorithm of Networked Control Systems. Ph.D Thesis, Zhejiang University, 2005.

[LI06] S.B. Li, Y.Q. Wang, F. Xia, Y.X. Sun. Guaranteed Cost Control of Networked Control Systems with Time-delays and Packet Losses. *International Journal of Wavelets, Multiresolution and Information Processing*, Vol. 4, No.4, pp.691 – 706, 2006.

[LIA01] F.L. Lian. Analysis, Design, Modeling, and Control of Networked Control Systems. Ph.D Thesis, University of Michigan, 2001.

[LIN02] B. Lincoln. Jitter Compensation in Digital Control Systems. *Proc. of the American Control Conf.*, 2002.

[LIN03a] Q. Ling, M.D. Lemmon. Soft Real-Time Scheduling of Networked Control Systems with Dropouts Governed by a Markov Chain. *American Control Conf.*, Denver, Colorado, 2003.

[LIN03b] Q. Ling, M.D. Lemmon. Optimal Dropout Compensation in Networked Control Systems. *Proc. of the IEEE Conf. on Decision and Control*, 2003.

[LIU73] C. Liu, J. Layland. Scheduling Algorithms for Multiprogramming in a Hard Real-Time Environment. *Journal of the ACM*, Vol.20, pp.46 – 61, 1973.

[LIU00] J.W.S. Liu. *Real-Time Systems*. Prentice Hall, 2000.

[LOK92] C.D. Locke. Software architecture for Hard Real-Time Applications: Cyclic vs. Fixed Priority Executives. *Real-Time Systems*, Vol. 4, pp. 37 – 53, 1992.

[MAR02] P. Martí. Analysis and Design of Real-Time Control Systems with Varying Control Timing Constraints. Ph.D Thesis, Technical University of Catalonia, 2002.

[MON04] L.A. Montestruque, P. Antsaklis. Stability of Model-based Networked Control Systems with Time-Varying Transmission Times. *IEEE Trans. Automatic Control*, Vol.49, No.9, pp.1562 – 1572, 2004.

[MUR03] R.M. Murray (Editor). Control in an Information Rich World: Report of the Panel on Future Directions in Control, Dynamics, and Systems. *Society for Industrial and Applied Mathematic*, 2003.

[NIL98] J. Nilsson. Real-Time Control Systems with Delays. Ph.D Thesis, Lund University, 1998.

[OHL06] M. Ohlin, D. Henriksson, A. Cervin. TrueTime 1.4-Reference Manual, Manual. *Department of Automatic Control*, Lund University, Sweden, 2006.

[PAL02] L. Palopoli. Design of Embedded Control Systems under Real-Time Scheduling Constraints. Ph. D Thesis, Scuola Superiore S. Anna, 2002.

[SAN00] M. Sanfridson. Timing Problems in Distributed Real-Time Computer Control Systems. *Technical Report*, Department of Machine Design, Royal Institute of Technology (KTH), 2000.

[SAN04] M. Sanfridson. Quality of Control and Real-Time Scheduling. Ph.D Thesis, Royal Institute of Technology (KTH), 2004.

[SCH06] L. Schenato. Optimal Estimation in Networked Control Systems Subject to Random Delay and Packet Loss. *IEEE Conf. on Decision and Control (CDC 06)*, San Diego, CA, USA, 2006.

[SHA04] L. Sha, T. Abdelzaher, K.E. Årzén, T. Baker, A. Burns, G. Buttazzo, M. Caccamo, A. Cervin, J. Lehoczky, A. Mok. Real-Time Scheduling Theory: A Historical Perspective. *Real-Time Systems*, Vol.28, pp. 101 – 155, 2004.

[STA96] J.A. Stankovic. Real-Time and Embedded Systems. *ACM Computing Surveys*, Vol. 28, No.1, pp.205 – 208, 1996.

[TAN02] A.S. Tanenbaum. *Computer Networks, 4th Edition*. Prentice Hall, PTR, 2002.

[TIA06] Y.C. Tian, D. Levy, M.O. Tadé, T. Gu, C. Fidge. Queuing Packets in Communication Networks for Networked Control Systems. *The 6th World Congress on Intelligent Control and Automation*, Dalian, China, pp. 205 – 209, 2006.

[TIP03] Y. Tipsuwan, M.Y. Chow. Control Methodologies in Networked Control Systems. *Control Engineering Practice*, Vol.11, pp.1099 – 1111, 2003.

[TOV99] E. Tovar. Supporting Real-Time Communications with Standard Factory-Floor Networks. Ph.D Thesis, University of Porto, 1999.

[WHE93] J.D. Wheelis. Process Control Communications: Token Bus. *CSMA/CD*, *or Token Ring*, *ISA Trans.*, Vol. 32, No. 2, pp.193 – 198, 1993.

[WIR07] Wireless Overview. The MAC level. http://www.hpl.hp.com/personal/Jean_Tourrilhes/Linux/Linux.Wireless.mac.html, 2007.

[WIT95] B. Wittenmark, J. Nilsson, M. Törngen. Timing Problems in Real-Time Control Systems. *Proc. of the American Control Conf.*, Seattle, WA, pp.2000 – 2004, 1995.

[WIT02] B. Wittenmark, K.J. Åström, K.E. Årzén. Computer Control: An Overview. *IFAC Professional Brief*, 2002.

[XIA06a] F. Xia. Feedback Scheduling of Real-Time Control Systems with Resource Constraints. Ph.D Thesis, Zhejiang University, 2006.

[XIA06b] F. Xia, Y.X. Sun. Control-Scheduling Codesign: A Perspective on Integrating Control and Computing. *Dynamics of Continuous, Discrete and Impulsive Systems-Series B: Applications and Algorithms, Special Issue on ICSCA'06*, Watam Press, pp.1352 – 1358, 2006.

[ZHA01] W. Zhang. Stability Analysis of Networked Control Systems. Ph.D Thesis, Department of Electrical Engineering and Computer Science, Case Western Reserve University, 2001.

[ZHO05] P. Zhou, J. Xie. Online Deadline Assignment to Reduce Output Jitter of Real-Time Control Tasks. *Proc. of Int. Conf. on Machine Learning and Cybernetics*, Vol. 2, pp.1312 – 1315, 2005.

PART II
CPU SCHEDULING

Chapter 3

Neural Feedback Scheduling

Optimal feedback scheduling schemes are usually too computationally expensive to be used online, though they are in principle capable of maximizing the overall QoC of Real-Time Control Systems. With the goal of optimizing the overall QoC of multitasking embedded control systems, the problem of optimal feedback scheduling is explicitly formulated, and relevant mathematical solutions are discussed.

To overcome the disadvantage of overly large computational overheads associated with mathematical optimization routines, a fast feedback scheduling scheme exploiting feedforward neural networks is presented. It aims at reducing the feedback scheduling overhead while delivering almost optimal overall QoC. Besides, the use of neural networks potentially enhances the adaptability, robustness, and fault-tolerance of the feedback scheduler. Simulation results argue that neural feedback scheduling can dramatically reduce the feedback scheduling overhead, while yielding overall QoC almost identical with optimal feedback scheduling.

3.1 Introduction

In multitasking embedded control systems, the most straightforward objective of feedback scheduling is to optimize the overall QoC. With this objective, the problem of feedback scheduling can naturally be formulated as a constrained optimization problem in which the total control cost is to be minimized through optimizing scheduling parameters under the constraint on system schedulability. This formulation constitutes the most common type of feedback scheduling problems, usually called optimal feedback scheduling. In the literature, the most popular solution for this problem is based on some mathematical optimization algorithms, e.g. [CER06, EKE00, HEN05, JIN05, SET96].

In order to deal with the dynamic variations of workload, as mentioned in Chapter 2, feedback schedulers arc usually executed online. Therefore, it is of paramount importance to take into account in practice the amount of resources consumed by the feedback scheduler itself, i.e. the computational complexity of the feedback scheduling algorithm.

If the feedback scheduler expends too much computing resource that is originally scarce, the execution of control tasks will inevitably be impacted negatively. This could then cause significant degradation of the overall QoC [ARZ03, HEN05, MAR02]. In this context the efficiency of the feedback scheduler becomes doubtful. While in theory optimal feedback scheduling schemes are effective in optimizing the overall QoC, typical algorithmic solutions for the optimal feedback scheduling problem often involve complex computations, which cause too large feedback scheduling overheads. Therefore, they are not suitable for online use in most cases.

To tackle the problem of the high computational complexity of optimal feedback scheduling algorithms, a neural network based fast, intelligent feedback scheduling scheme [XIA05, XIA07] is presented in this chapter, namely neural feedback scheduling (NFS). As a widely used intelligent computing technique, neural networks have powerful capability of approximating arbitrary nonlinear functions with arbitrary precision [SUN97, SUN06, YAN04]. Consequently, neural feedback scheduling can yield almost optimal solutions. On the other hand, since feedforward BP (Back Propagation) neural networks with a simple structure are adopted here, the online computational complexity will be reduced in comparison with the optimal feedback scheduler. The highly parallel processing mechanism of neural networks also makes it possible to implement much faster feedback schedulers by hardware. In addition, the application of neural networks naturally brings intelligent characteristics to the feedback scheduler. As a result, its adaptability, robustness, and fault-tolerance in dynamic and unpredictable environments will be reinforced.

With the same goal of reducing the feedback scheduling overhead, Cervin, *et al.* [CER02] present a simple rescaling based feedback-feedforward scheduling scheme to approximate optimal solutions. It has proved quite effective in some cases where the control cost functions can be approximately described by either quadratic or linear functions of sampling periods. However, the performance of this simple rescaling scheme is guaranteed only for systems with cost functions in some specific forms. Therefore, it may not always be applicable due to considerably worse-than-optimal performance. In contrast, the neural network based solution presented in this chapter is widely applicable.

The major contributions of this chapter are as follows:

(1) The problem of optimal feedback scheduling is explicitly formulated and related mathematical solutions are discussed.

(2) A novel feedback scheduling scheme, i. e. neural feedback scheduling, is presented, with exciting advantages such as low feedback scheduling overhead, wide applicability, being intelligent, etc. It is also capable of delivering almost optimal overall QoC.

(3) For the area of neural networks, a novel and successful application is demonstrated. Meanwhile, from the perspective of feedback scheduling, a completely fresh intelligent solution is suggested.

The rest of this chapter is organized as follows. Within the feedback scheduling framework given in Chapter 2, Section 3.2 first formulates the problem of optimal feedback scheduling, and then discusses related solutions that exploit mathematical optimization routines. After the major disadvantage of optimal feedback scheduling algorithms is pointed out, the neural feedback scheduling scheme is described in Section 3.3, followed by the analysis of its computational complexity. Section 3.4 assesses the performance of neural feedback scheduling via simulation experiments. It is also compared against optimal feedback scheduling. Finally, this chapter is concluded in Section 3.5.

3.2 Optimal Feedback Scheduling

In this section, the optimal feedback scheduling problem is described. The high computational complexity of typical mathematical/algorithmic solutions for this optimization problem is discussed. Thus the motivation for this work will be further illustrated. In addition, the formulation and solution of the optimal feedback scheduling problem will serve as an essential of the design of the neural feedback scheduler, see Section 3.3.

3.2.1 Problem Formulation

Consider a multitasking embedded control system as shown in Fig.3.1, where N independent control tasks share a processor with limited processing capability. Besides control tasks, other (high-priority) non-control tasks may exist on the same processor. The timing attributes of these non-control tasks cannot be altered on purpose. In this monograph the terms *non-control task* and *disturbing task* are used interchangeably wherever possible. The execution times of control tasks and the requested CPU utilization of non-control tasks may change over time.

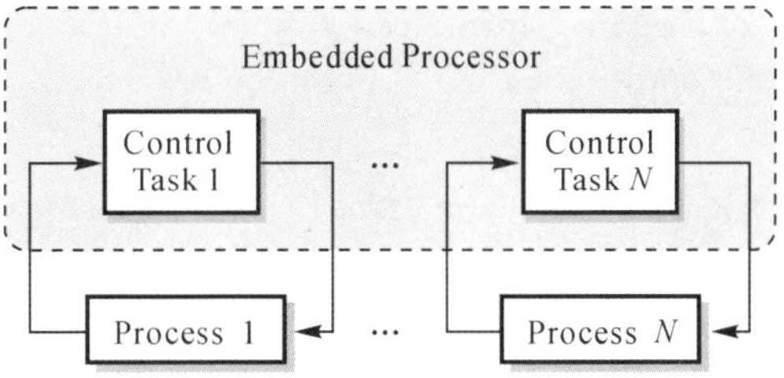

Fig. 3.1 A multitasking embedded control system

The goal is to optimally allocate available CPU resources among control loops, and ultimately to optimize the overall quality of the control of the system, by exploiting feedback scheduling. In the feedback scheduling loop, the CPU utilization is chosen as the controlled variable. It is the responsibility of the feedback scheduler to adapt sampling periods (i. e. task periods) so that the CPU utilization is maintained at a desired level. For simplicity, assume that all task execution times and the CPU workload are available at runtime.

According to sampled-data control theory, smaller sampling periods yield better control performance, because the performance of sampled-data control with smaller sampling periods approaches more closely to that of continuous-time control. However, the decrease in sampling period will adversely increase the requested CPU utilization of the corresponding control task. In extreme cases the schedulability of the system may be violated, and hence the control performance will deteriorate due to deadline misses. In order to optimize the overall QoC, the sampling periods should be adjusted under the constraint on system schedulability. From the resource management perspective, the available computing resources should be distributed among control tasks in an optimal way that the total control cost of the system is minimized.

Let h_i and c_i denote the (sampling) period and the execution time of control task i respectively. The problem of optimal feedback scheduling can be formulated as [EKE00, SET96]:

$$\begin{aligned} \min_{h_1, \cdots, h_N} J &= \sum_{i=1}^{N} w_i J_i(h_i) \\ \text{s. t.} \sum_{i=1}^{N} \frac{c_i}{h_i} &\leqslant U_R \end{aligned} , \tag{3.1}$$

where $J_i(h_i)$ is the control performance index (i.e. cost function) of loop i, as a function of the sampling period, w_i is a weight reflecting the relative importance of each loop, U_R is the allowable utilization upper bound for all control tasks, which is closely related to the underlying scheduling policy employed and the requested utilization of disturbing tasks. In general, $J_i(h_i)$ is monotonically increasing [ARZ99, AST97].

The optimal feedback scheduling problem described by Eq.(3.1) is a typical constrained optimization problem. To obtain linear constraints, the costs are generally recast as functions of sampling frequencies $f_i = 1/h_i$ instead of periods [CER02, EKE00]. By argument substitution, Eq.(3.1) can be rewritten as:

$$\begin{aligned} \min_{f_1, \cdots, f_N} J &= \sum_{i=1}^{N} w_i J_i(f_i) \\ \text{s. t.} \sum_{i=1}^{N} c_i f_i &\leqslant U_R \end{aligned} . \tag{3.2}$$

In the above formulation, it is important to choose an appropriate cost function. According to the principle of LQ optimal control, a linear quadratic cost function may be chosen. For example, [CER06, EKE00, HEN05, JIN05] use such a form of cost functions. In [CER02], Cervin, *et al.* present two approximate performance indexes as given in Eqs.(1.3) and (1.4). Another type of approximate cost function is exponential functions, e.g. [BRA02, HE04, KIM98, SET96].

In most of these cost functions it is common to neglect the impact of delay on control performance. Since this chapter focuses on the effect of sampling periods rather than time delay, the influence of delay and its jitter is disregarded when solving the optimal feedback scheduling problem. It is worth noting that delay and its jitter do affect control performance in our simulation experiments. In other words, although delay and its jitter are excluded from the formulation of cost functions, their impact on QoC will be examined via simulation experiments.

Also note that the neural feedback scheduling scheme developed in this chapter does not rely on control cost functions of any specific forms. It is applicable to control systems with arbitrary cost functions provided that Eq.(3.2) can be solved offline.

3.2.2 Mathematical Optimization Methods

In the area of optimization, there exist quite a lot of well-established algorithms for solving the constrained optimization problem formulated by Eq.(3.2). Generally speaking, the majority of these methods could be divided into two categories [BOY04, SUN04]. The first directly applies some searching methods to the objective function so as to obtain the optimal solutions on feasible directions, e.g. the feasible direction method. The second attempts to transform the original complex problem into simpler ones that are much easier to solve and solves these simple problems one by one. The penalty function method and quadratic programming are two examples of the second kind.

Given below is the necessary and sufficient condition for the optimal solutions of Eq.(3.2).

Theorem 3.1 (**Kuhn-Tucker condition**) [BOY04, SUN04] For the optimal feedback scheduling problem formulated by Eq.(3.2), given that $J_i(f_i)$ is convex, $\boldsymbol{f}' = [f_1', \cdots, f_N']^{\mathrm{T}}$ is the optimal solution if and only if

$$\begin{aligned} &\nabla \boldsymbol{J}(\boldsymbol{f}') + \lambda \boldsymbol{c} = 0 \\ &\lambda [U_R - \boldsymbol{c}^{\mathrm{T}} \boldsymbol{f}'] = 0 , \\ &\lambda \geqslant 0 \end{aligned} \tag{3.3}$$

where $\nabla \boldsymbol{J}$ is the gradient vector, $\boldsymbol{c} = [c_1, \cdots, c_N]^{\mathrm{T}}$, and λ is the Lagrange multiplier.

The above Kuhn-Tucker condition is one of the most important results in the field of nonlinear programming. Solving Eq.(3.3) is the basis of a lot of nonlinear programming methods.

In cases where $J_i(f_i)$ is not convex, Eq.(3.3) will become only a necessary condition for solving the optimal feedback scheduling problem. Thanks to the fact that $J_i(f_i)$ is indeed convex in most control systems, the Kuhn-Tucker condition can be regarded as a general tool for obtaining the optimal sampling frequencies/periods.

Besides many others, sequential quadratic programming (SQP) [FLE87] has been recognized as one of the most efficient methods for solving constrained optimization problems. It can emulate the way that Newton's method solves unconstrained optimization problems. In each iteration, it constructs a quadratic programming sub-problem, and uses the solution for this sub-problem as the direction of search, thus approaching the final solution for the constrained optimization problem. The convergence of the algorithm is guaranteed by the Hessian matrix composed of Lagrange functions obtained using quasi-Newton methods. For the sake of simplicity, it is assumed hereafter that the SQP method is by default used for the optimal feedback scheduling scheme wherever it is involved.

It is not difficult to find out that complex computations associated with gradients and Hessian matrices are involved in solving Eq.(3.3) using mathematical optimization algorithms, and that quite a large number of iterations are usually required before reaching the final solution. This makes the computational complexity of the algorithms considerably high. In the SQP method, for instance, one or two quadratic programming sub-problems must be solved during each iteration, which could take a considerable amount of time to complete. When applied to feedback scheduling, this algorithm may cause a large feedback scheduling overhead consuming too much computing resource. For this reason, optimal feedback schedulers based on mathematical optimization methods are usually not suitable for online use.

In fact, almost all existing optimal feedback scheduling algorithms such as [CER06, EKE00, HEN05] suffer from the problem of too large computational overheads, which impair the practicability of these algorithms. In particular, when $J_i(f_i)$ is not convex, online optimization routines will become extraordinarily complex and time-consuming. Consequently, the optimal feedback scheduler will not be able to perform as expected. In some cases, it may even cause degradation of the system performance.

3.3 Neural Feedback Scheduling Scheme

Since optimal feedback scheduling schemes are generally too computationally expensive to be used online, schemes with much less computational complexity are desirable. In this section, an efficient feedback scheduling scheme using neural networks will be presented. Some reasons why the neural networks technology is introduced into the framework of feedback scheduling are described below:

(1) From the overhead perspective, feedforward neural networks with simple structures can yield smaller feedback scheduling overheads than mathematical optimization methods thanks to less complex computations. In addition, information processing inside neural networks is highly parallel, which makes it possible to achieve higher processing speed by means of suitable hardware implementation.

(2) With regard to the accuracy of solutions, mathematical optimization methods can generate accurate optimal solutions offline, which should be exploited in the design of online feedback schedulers. On the other hand, neural networks are powerful in learning and adapting, and capable of approximating arbitrarily complex nonlinear functions with arbitrary precision. Once well trained by the accurate optimal solutions at design time, neural networks will be able to deliver online almost optimal feedback scheduling performance.

(3) The generalization capability of neural networks is also very good in that neural networks can easily handle things like untrained input data, noise, and incomplete data. This helps to improve the performance of the feedback scheduler with better fault tolerance.

3.3.1 Design Methodology

The basic idea behind neural feedback scheduling is to use a feedforward neural network to approximate the optimal solutions, which are obtained using mathematical optimization methods, e.g. SQP. Following this idea, training and testing data will not be a problem since it can be easily created offline by applying the SQP method to the corresponding optimal feedback scheduling problem. After the procedure of pre-processing (e.g. normalization), data samples will be used to train and test the neural networks. In the following the structure of the neural feedback scheduler and the design flow will be described.

This chapter selects a three-layer feedforward BP network to constitute the feedback scheduler. Two major reasons for the choice of a BP network are given below:

(1) The structure of BP neural networks is simple, which is beneficial to simplifying online computations. Additionally, BP networks are easy to implement.

(2) BP networks are the most widely used neural network technology applied in practice, while exhibiting the main characteristics of neural networks. It has been reported that in real-world applications 80% – 90% neural network models are based on BP networks and variants [SU03, YAN04].

As shown in Fig.3.2, there is only one hidden layer apart from the input and output layers in the BP network used as the feedback scheduler. Existing theoretical analysis argues that feedforward neural networks with only one hidden layer are able to map arbitrary functions with arbitrary precision that are continuous on closed intervals [SUN97, YAN04]. Therefore, one hidden layer is sufficient for guaranteeing solution accuracy.

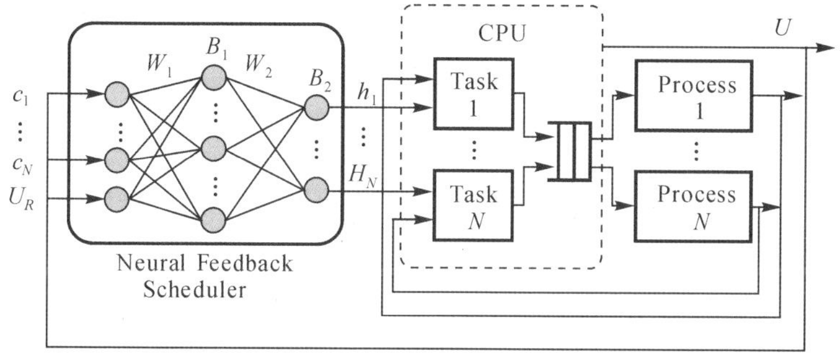

Fig. 3.2 Architecture of neural feedback scheduling

According to Eq.(3.2), with given cost functions, the values of the sampling frequencies will depend on the execution time c_i of each control task and the desired CPU utilization U_R. As a consequence of this observation, $(N+1)$ inputs, i.e. c_1, $\cdots$, c_N, U_R, are set for the neural feedback scheduler. Since the role of the feedback scheduler is to determine sampling periods of all loops, the sampling periods (h_1, $\cdots$, h_N) or frequencies (f_1, $\cdots$, f_N) are natural outputs of the neural feedback scheduler.

From a real-time scheduling point of view, both inputs and outputs of the feedback scheduler are associated with resource utilization. The role of the neural feedback scheduler can also be explained as dynamically allocating the available computing resources. From a control perspective, sampling periods/frequencies are important design parameters of control loops. Therefore, the neural feedback scheduler establishes a mapping from the temporal parameters (for real-time scheduling) to the controller parameters (for real-time control).

Using the neural network structure given in Fig.3.2, the relationship between the outputs and the inputs of the neural feedback scheduler is expressed as:

$$\boldsymbol{Y} = \boldsymbol{W}_2[\sigma(\boldsymbol{W}_1\boldsymbol{X} + \boldsymbol{B}_1)] + \boldsymbol{B}_2, \tag{3.4}$$

where $\boldsymbol{W}_1$, $\boldsymbol{W}_2$, $\boldsymbol{B}_1$, and $\boldsymbol{B}_2$ are weight matrices and bias vectors respectively, the input vector $\boldsymbol{X} = [c_1, \cdots, c_N, U_R]^{\mathrm{T}}$, and the output vector $\boldsymbol{Y} = [f_1, \cdots, f_N]^{\mathrm{T}}$ (or $[h_1, \cdots, h_N]^{\mathrm{T}}$). The activation functions used are the sigmoid transfer function $\sigma(x) = 1/1 + \mathrm{e}^{-x}$ in the hidden layer and the linear transfer function in the output layer.

In designing a neural feedback scheduler, it is of great importance to determine an appropriate number of neurons in the hidden layer (i.e. hidden neurons). While some instructions exist in the neural networks field, there is no general theory for determining the number of hidden neurons. Therefore, practical experience and simulation studies must be relied on in most cases.

As Eq.(3.4) indicates, the number of hidden neurons is tightly associated with the computational complexity of the neural feedback scheduler. In the case of too many hidden neurons, the feedback scheduler will consume too much computing resource, thus causing large feedback scheduling overheads. Fortunately, the number of control tasks that run concurrently on the same CPU is usually limited in practice. Accordingly, it is not necessary in most cases for the number of hidden neurons to be very large. This ensures that the computations associated with the neural feedback scheduler will not be overly time-consuming.

In the training of the BP network, the Levengerg-Marquardt algorithm is adopted, which can not only improve the performance of traditional BP networks but also speed the convergence of network training [SU03].

As Fig.3.3 shows, the design flow of the neural feedback scheduler is as follows. First, formulate the problem in the form of constrained optimization as given by Eq.(3.2), determine the form of cost functions based on control systems analysis, and initialize related parameters. Second, analyze the characteristics of the execution times of control tasks obtaining the ranges of their values. Within the ranges of c_i as well as U_R, select a number of data pairs, and for each pair, use the SQP method to solve the optimal feedback scheduling problem offline, thus producing sufficient sample datasets. Third, determine the number of hidden neurons according to the number of control loops and initialize the neural network. Then, train and test the neural network using pre-processed sample datasets. Once the BP network passes the test, it can thereafter be used online as the neural feedback scheduler.

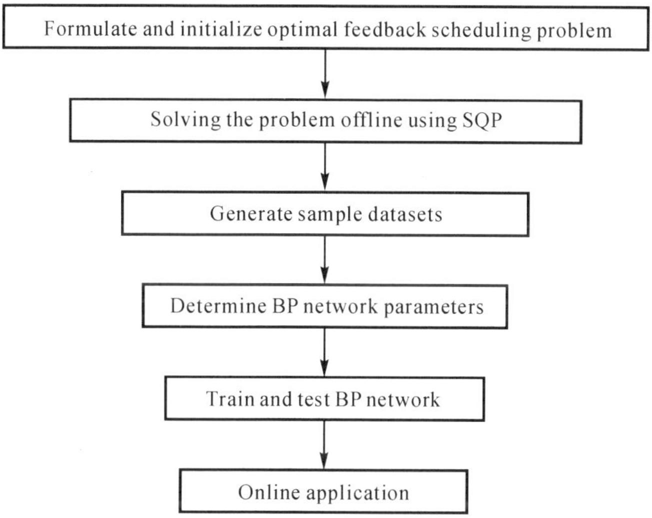

Fig. 3. 3 Design flow of the neural feedback scheduler

The online application of the well-designed neural feedback scheduler is quite simple. At every invocation instant, the feedback scheduler gathers the current values of all input variables, and then calculates the sampling periods/frequencies using Eq.(3.4), followed by the update of the sampling period of each loop.

3. 3. 2 Complexity Analysis

The computational complexity of the neural network based feedback scheduler presented above is analyzed below, with comparison against the SQP method. Though the application of neural feedback schedulers involves not only online computations but also offline computations, e.g. mathematical optimization, network training and test, etc., only the complexity of online computations is of concern, because it is the computational operations at runtime that decide the feedback scheduling overhead.

With a given invocation interval, the amount of computing resources consumed by the feedback scheduler depends directly on the CPU time needed for each run. To analyze the computational complexity of the neural feedback scheduler, consider again the online calculation Eq.(3.4). Let M be the number of hidden neurons. Eq.(3.4) can easily be decomposed into the following three sub-equations:

$$\begin{cases} \boldsymbol{A} = \boldsymbol{W}_1\boldsymbol{X} + \boldsymbol{B}_1 \\ \boldsymbol{Z} = \sigma(\boldsymbol{A}) \\ \boldsymbol{Y} = \boldsymbol{W}_2\boldsymbol{Z} + \boldsymbol{B}_2 \end{cases}, \tag{3.5}$$

where $\boldsymbol{A} = [a_1, \cdots, a_M]^{\mathrm{T}}$ and $\boldsymbol{Z} = [z_1, \cdots, z_M]^{\mathrm{T}}$ are temporary arguments (vectors),

$$\boldsymbol{W}_1 = \begin{bmatrix} w'_{11} & \cdots & w'_{1,N+1} \\ \vdots & \ddots & \vdots \\ w'_{M,1} & \cdots & w'_{M,N+1} \end{bmatrix}, \boldsymbol{W}_2 = \begin{bmatrix} w''_{11} & \cdots & w''_{1,M} \\ \vdots & \ddots & \vdots \\ w''_{N,1} & \cdots & w''_{N,M} \end{bmatrix},$$

$\boldsymbol{B}_1 = [b'_1, \cdots, b'_M]^{\mathrm{T}}$, and $\boldsymbol{B}_2 = [b''_1, \cdots, b''_N]$.

Eq.(3.5) can be rewritten as:

$$\begin{bmatrix} a_1 \\ \vdots \\ a_{M-1} \\ a_M \end{bmatrix} = \begin{bmatrix} w'_{11} & \cdots & w'_{1,N} & w'_{1,N+1} \\ \vdots & \ddots & & \vdots \\ & w'_{M-1,1} & w'_{M-1,N} & w'_{M-1,N+1} \\ w'_{M,1} & \cdots & w'_{M,N} & w'_{M,N+1} \end{bmatrix} \begin{bmatrix} c_1 \\ \vdots \\ c_N \\ U_R \end{bmatrix} + \begin{bmatrix} b'_1 \\ \vdots \\ b'_N \\ b'_{N+1} \end{bmatrix}, \tag{3.6}$$

$$z_i = \sigma(a_i) = \frac{1}{1 + \mathrm{e}^{-a_i}}, \quad i = 1, \cdots, M, \tag{3.7}$$

$$\begin{bmatrix} f_1 \\ \vdots \\ f_N \end{bmatrix} = \begin{bmatrix} w''_{11} & \cdots & w''_{1,M} \\ \vdots & \ddots & \vdots \\ w''_{N,1} & \cdots & w''_{N,M} \end{bmatrix} \begin{bmatrix} z_1 \\ \vdots \\ z_M \end{bmatrix} + \begin{bmatrix} b''_1 \\ \vdots \\ b''_N \end{bmatrix}. \tag{3.8}$$

The above three equations give almost all operations that the neural feedback scheduler has to complete each time it is invoked. The number of basic operations associated with these computations is roughly summarized in Table 3.1.

As can be seen from Table 3.1, there are $(4MN + 6M - N)$ basic operations in each run of the neural feedback scheduler. Clearly, the feedback scheduling overhead relates primarily to the number of control loops and hidden neurons, i.e. N and M. In fact, in almost all cases the number of hidden neurons is chosen according to the number of network inputs, which are determined by the number of concurrent control tasks. As a consequence, the number of control tasks can be regarded as the most important factor that determines the overhead of neural feedback scheduling.

In general, M is proportional to N, i.e. $M \propto N$ (e.g. $M \approx 2N$). Therefore, the time complexity of the neural feedback scheduling algorithm is $O(N^2)$. Meanwhile, the computational complexity of a typical mathematical optimization algorithm, e.g. SQP, is (at least) $O(N^3)$ [CAS06]. From this insight, neural feedback scheduling reduced the computational complexity of the algorithm in comparison with optimal feedback scheduling.

As mentioned above, the value of N is always limited in real systems. Consequently, a more convincing method for examining the runtime efficiency of feedback schedulers is to compare the actual CPU time consumed by different feedback schedulers via simulations and/or real experiments, see Section 3.4.

Table 3.1 Complexity analysis of neural feedback scheduling

	Addition (Subtraction)	Multiplication	Division	Power
Eq. (3.6)	$M(N+1)$	$M(N+1)$	0	0
Eq. (3.7)	M	0	$2M$	M
Eq. (3.8)	MN	$(M-1)N$	0	0
Total	$2M(N+1)$	$2MN+M-N$	$2M$	M

3.4 Performance Evaluation

This section will test and analyze the performance of the proposed neural feedback scheduling scheme via simulation experiments. From a control perspective, the purpose of this evaluation is twofold. The first is to validate the effectiveness of neural feedback scheduling, i.e. to check whether or not it is able to deal with dynamic variations in both the control tasks' resource demands and the available resources. The second is to study the difference between neural feedback scheduling and ideal optimal feedback scheduling in terms of overall QoC. From the viewpoint of implementation efficiency, the actual time overheads of different feedback schedulers will be compared, thus highlighting the major merit of the neural feedback scheduling scheme.

3.4.1 Setup Overview

Consider a multitasking embedded processor that is responsible for controlling three inverted pendulums concurrently, as shown in Fig.3.4. Accordingly, there are three independent control tasks. The linearized state-space models of the inverted pendulums are in the following form [CER02]:

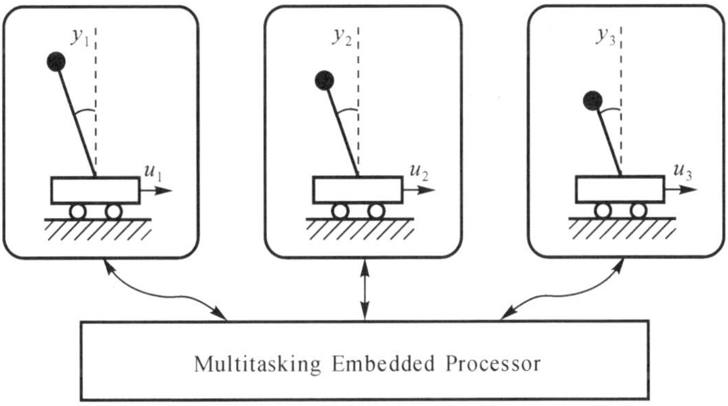

Fig. 3.4 The simulated system

$$\begin{aligned} \dot{\boldsymbol{x}}(t) &= \begin{bmatrix} 0 & 1 \\ \omega_0^2 & 0 \end{bmatrix} \boldsymbol{x}(t) + \begin{bmatrix} 0 \\ \omega_0^2 \end{bmatrix} \boldsymbol{u}(t) + \boldsymbol{v}(t) \\ \boldsymbol{y}(t) &= \begin{bmatrix} 1 & 0 \end{bmatrix} \boldsymbol{x}(t) + \boldsymbol{e}(t) \end{aligned} , \tag{3.9}$$

where ω_0 is the natural frequency of the inverted pendulum, $\boldsymbol{v}$ and $\boldsymbol{e}$ are sequences of white Gaussian noise with zero mean and variances of $1/\omega_0$ and 10^{-4}, respectively.

Due to the difference in length, the three inverted pendulums have different natural frequencies given by $\omega_0 = 10$, 13.3, and 16.6, respectively. All initial states are zero. Every pendulum is controlled independently by a LQG (Linear Quadratic Gaussian) controller, whose objective is to minimize the following cost function:

$$J = \int_0^{\infty} (y^2 + u^2)\,\mathrm{d}t . \tag{3.10}$$

For the sake of simplicity, the approximate performance index given below is used for Eq.(3.2) in simulations [CER02]:

$$J_i(f_i) = \alpha_i + \frac{\gamma_i}{f_i}, \tag{3.11}$$

where $\gamma_i = 43$, 67, and 95, respectively, for each control loop.

The initial sampling frequency of each loop is chosen as $f_0 = 58.8$, 71.4, and 83.3 Hz, respectively. Also, assume $w_i = 1$ for simplicity.

Besides these three control tasks, there is an additional periodic non-control task running on the same processor. The execution time of this task is variable, causing U_R to vary over time. Disregarding the execution of the feedback scheduling task, the desired total CPU utilization of all tasks is set to $\widetilde{U}_R = 0.75 < 4(2^{1/4} - 1) = 0.76$. According to Theorem 2.1, the system schedulability under the RM algorithm is guaranteed by $\widetilde{U}_R$. The execution time of the non-control task is c_4, and its period $h_4 = 10$ ms. Therefore, $U_R = \widetilde{U}_R - c_4/h_4$, implying that U_R will change with c_4.

Task priorities in the system are assigned as follows. The feedback scheduling task has the highest priority, and the priorities of other tasks are determined in accordance with the RM policy. The invocation interval of feedback scheduler is $T_{\mathrm{FS}} = 400$ ms.

To measure the overall QoC, the total control cost J_{SUM} of three control loops is recorded:

$$J_{\mathrm{SUM}}(t) = \sum_{i=1}^{3} w_i J_i(t) = \sum_{i=1}^{3} \int_0^t [y^2(\tau) + u^2(\tau)]\,\mathrm{d}\tau . \tag{3.12}$$

Intuitively, the bigger the value of J_{SUM}, the worse the overall QoC.

3.4.2 Neural Feedback Scheduler Design

The neural feedback scheduler is designed following the procedures shown in Fig.3.3. Based on the above description of the simulated system, the following formulation of the corresponding optimal feedback scheduling problem is obtained:

$$\begin{aligned} \min_{f_1, f_2, f_2} J &= \alpha + \frac{43}{f_1} + \frac{67}{f_2} + \frac{95}{f_3} \\ \text{s. t. } c_1 f_1 + c_2 f_2 + c_3 f_3 &\leqslant 0.75 - \frac{c_4}{0.01} \end{aligned}, \tag{3.13}$$

where $\alpha = \alpha_1 + \alpha_2 + \alpha_3$ is a constant.

For the purpose of creating sample data, the ranges of c_1, c_2 and c_3 are set to [2, 9], [2, 7], and [1, 7], respectively, with increments of 1. c_4 takes on values ranging from 0.5 to 3 with increments of 0.5. The unit of these parameters is ms. For all possible values of these parameters, applying SQP to solve Eq.(3.13) offline results in totally 2,016 sets of sample data.

To further simplify online computations, the sampling periods instead of the frequencies are used as the outputs of the neural feedback scheduler. Some examples of the sample datasets are given in Table 3.2, where all timing parameters are given in ms.

Table 3.2 Examples of sample datasets

No.	c_1	c_2	c_3	U_R	h_1	h_2	h_3
1	0.002	0.002	0.001	0.70	0.0097	0.0073	0.0045
2	0.002	0.002	0.001	0.65	0.0104	0.0079	0.0049
3	0.002	0.002	0.001	0.60	0.0113	0.0085	0.0053
4	0.002	0.002	0.001	0.55	0.0123	0.0093	0.0058
5	0.002	0.002	0.001	0.50	0.0135	0.0103	0.0063
…							
501	0.003	0.007	0.007	0.60	0.0254	0.0318	0.0268
502	0.003	0.007	0.007	0.55	0.0282	0.0346	0.0290
503	0.003	0.007	0.007	0.50	0.0311	0.0380	0.0319
504	0.003	0.007	0.007	0.45	0.0345	0.0422	0.0355
505	0.004	0.002	0.001	0.70	0.0153	0.0076	0.0057
…							
1111	0.006	0.004	0.004	0.70	0.0277	0.0182	0.0151
1112	0.006	0.004	0.004	0.65	0.0298	0.0198	0.0162
1113	0.006	0.004	0.004	0.60	0.0323	0.0211	0.0178
1114	0.006	0.004	0.004	0.55	0.0353	0.0231	0.0194
1115	0.006	0.004	0.004	0.50	0.0388	0.0254	0.0213
…							
1626	0.008	0.004	0.005	0.45	0.0544	0.0308	0.0289
1627	0.008	0.004	0.006	0.7	0.0362	0.0206	0.0210
1628	0.008	0.004	0.006	0.65	0.0391	0.0219	0.0228
1629	0.008	0.004	0.006	0.6	0.0423	0.0239	0.0246
1630	0.008	0.004	0.006	0.55	0.0461	0.0261	0.0269
…							

Once the sample datasets are created, they will be normalized onto the interval [0, 1]. Since c_i and U_R are on rather different orders, normalizing original sample data can avoid saturations of neurons and speed the convergence of the neural network.

In order to determine the number of hidden neurons, i.e. the value of M, neural networks of different sizes are compared. Fig.3.5 gives the training errors of neural networks with M = 4, 6, 8, and 10, respectively. Given that the performance is comparable, a smaller M value should be chosen in order to reduce the feedback scheduling overhead. From this insight, it is set that M = 8 because of the good performance of corresponding neural network.

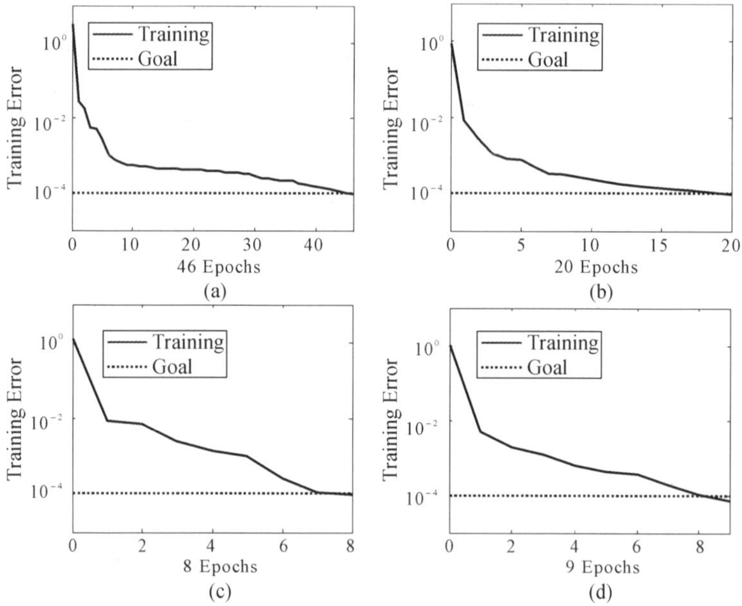

Fig. 3.5 Training errors for neural networks of different sizes
(a) $M = 4$; (b) $M = 6$; (c) $M = 8$; (d) $M = 10$

Remark 3. 1 It is not imperative in this chapter that the sample data be normalized, because all original data falls inside the interval [0, 1]. In general cases, however, normalization helps improve the performance of neural networks.

3.4.3 Results and Analysis

QoC

First, we examine the control performance of the system with different schemes. The execution time of each task varies at runtime according to Fig.3.6. The overhead of feedback scheduling is neglected here and will be studied later.

The following three schemes are compared in the simulations:

(1) Open-Loop Scheduling (OLS): All control loops use fixed (initial) sampling periods.

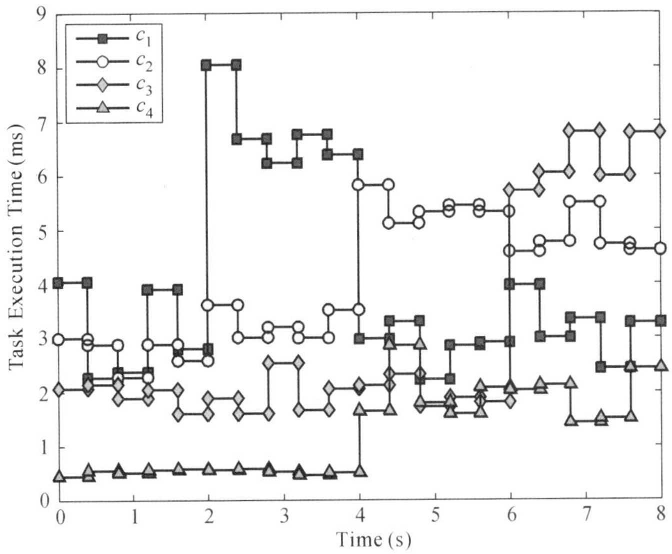

Fig. 3. 6 Task execution time

(2) Optimal Feedback Scheduling (OFS): The optimal feedback scheduling scheme that uses the SQP method. This is an *idealized* case for QoC optimization because the online computational overhead is assumed to be zero.

(3) Neural Feedback Scheduling (NFS): The method presented in this chapter.

Fig.3.7 depicts the total control cost of the system calculated using Eq.(3.12) when each of the three schemes is applied. With the traditional open-loop scheduling scheme, the system finally becomes unstable. Compared with the other two feedback scheduling schemes, the open-loop scheduling yields the worst overall QoC.

After time instant $t = 6$ s, as indicated in Fig.3.7, at least one pendulum falls down under open-loop scheduling. At $t = 6$ s, the task execution time $c_i = 0.004$, 0.0046, 0.0057, and 0.002 s, respectively. The total requested CPU utilization of all tasks is

$$U_{\text{req}} = \sum \frac{c_i}{h_i} = \frac{0.004}{0.017} + \frac{0.0046}{0.014} + \frac{0.0057}{0.012} + \frac{0.002}{0.01} = 1.24 > 1.$$

Obviously, the system is unschedulable. After this time instant, the total requested CPU utilization remains very high all along (see Fig.3.12), thereby leading to system instability. Furthermore, according to the principle of the RM algorithm, the priority of Task 1 is the lowest due to its largest period. Therefore, the first pendulum is finally out of control, as shown in Fig.3.8. In addition, control Loop 3 performs best thanks to its highest priority among all control tasks.

By comparing NFS with OLS via Fig.3.7, it can be found that NFS is quite effective in dealing with dynamic variations in both task execution time and available resources. The comparison of NFS and OFS, on the other hand, argues that NFS

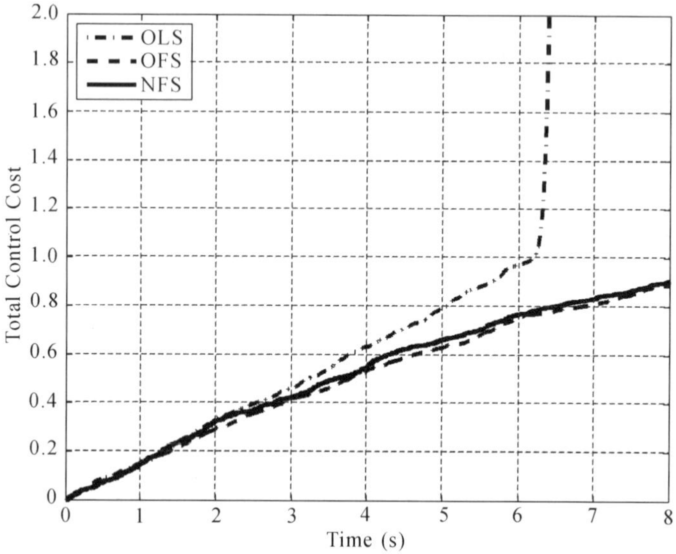

Fig. 3.7 Total control costs under different schemes

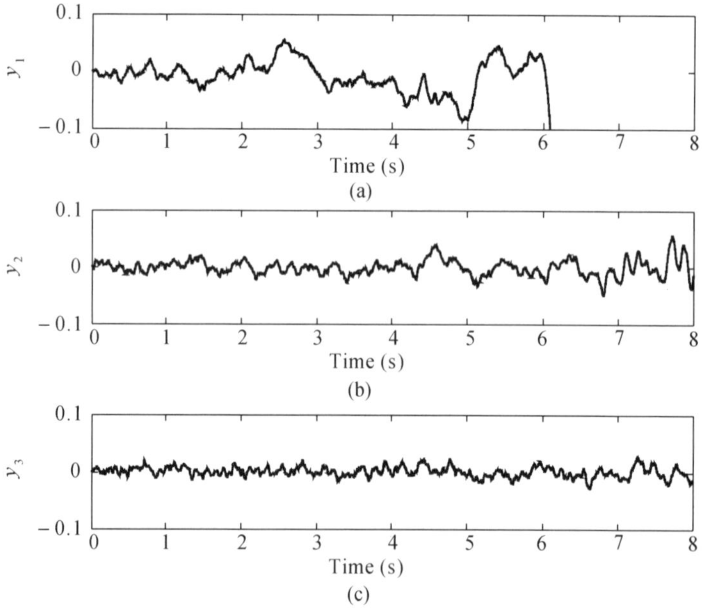

Fig. 3.8 System outputs under open-loop scheduling
(a) Inverted Pendulum 1; (b) Inverted Pendulum 2; (c) Inverted Pendulum 3

can deliver almost the same overall QoC as OFS. Figs.3.9 and 3.10 show the control performance of each inverted pendulum under OFS and NFS respectively.

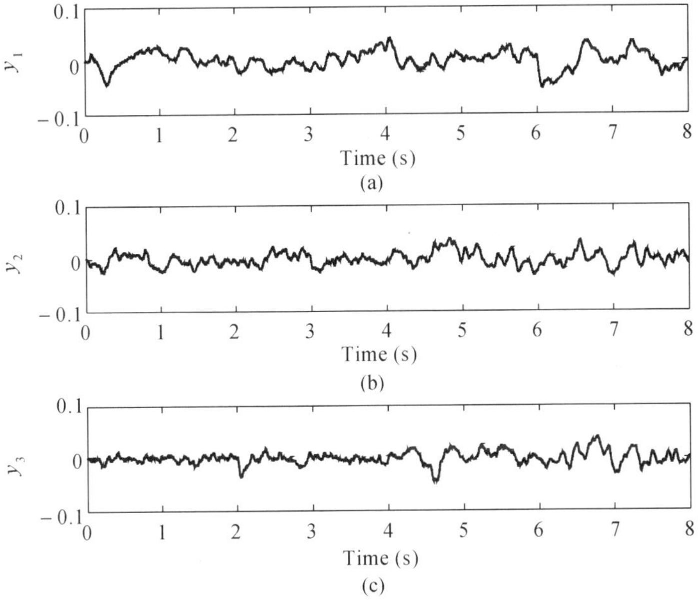

Fig. 3.9 System outputs under optimal feedback scheduling
(a) Inverted Pendulum 1; (b) Inverted Pendulum 2; (c) Inverted Pendulum 3

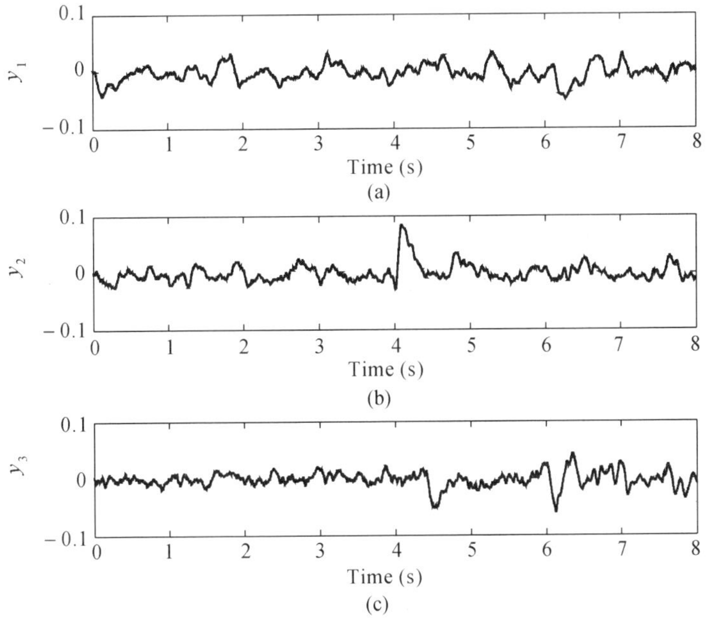

Fig. 3.10 System outputs under neural feedback scheduling
(a) Inverted Pendulum 1; (b) Inverted Pendulum 2; (c) Inverted Pendulum 3

Table 3.3 summarizes the accumulated control cost of each loop as well as the total control cost when different schemes are employed, where $J = \infty$ indicates that the system is unstable. It is clear that the overall QoC under NFS and OFS is almost the same, and is much better than that under OLS.

Table 3.3 Control costs under different schemes

Schemes	J_1	J_2	J_3	J_{SUM}
OLS	∞	0.424	0.356	∞
OFS	0.311	0.283	0.296	0.890
NFS	0.302	0.264	0.340	0.906

To examine the difference between NFS and OFS in more detail, Fig. 3.11 depicts the sampling periods of three control loops under different schemes. In contrast to the fixed sampling periods under OLS, both NFS and OFS adapt sampling periods at runtime. All sampling periods under NFS and OFS are nearly the same, though denoted by different types of lines in Fig. 3.11 for clear illustration. This validates that neural feedback scheduling is able to generate almost the same results as optimal feedback scheduling. The difference in the resulting control performance under these two schemes mainly derives from the stochastic noises inside the control systems.

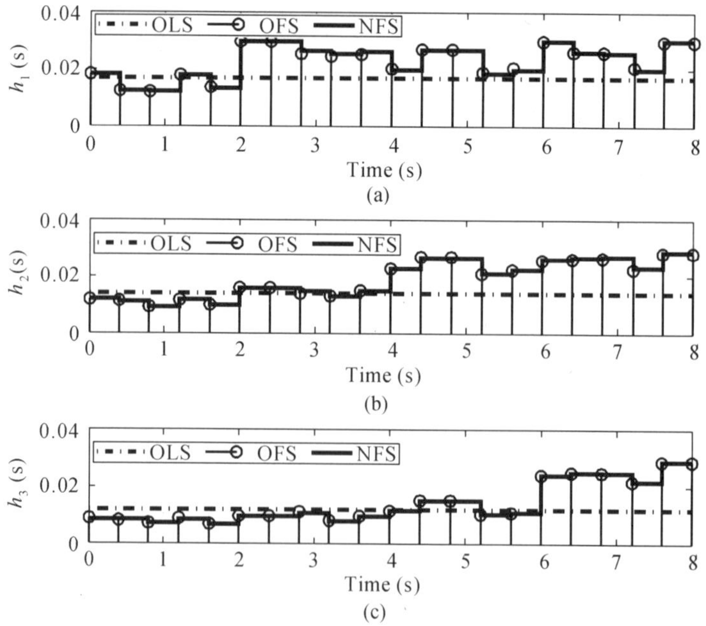

Fig. 3.11 Sampling periods under different schemes
(a) Inverted Pendulum 1; (b) Inverted Pendulum 2; (c) Inverted Pendulum 3

The adjustment of sampling periods by feedback schedulers results in changes in CPU workload, i. e. the total requested CPU utilization of all tasks U_{req}. The purpose is of course to achieve the desired CPU utilization.

As shown in Fig.3.12, when the open-loop scheduling scheme is used, the CPU workload changes with task execution times, because all task periods are fixed. After time instant t = 6 s, the total requested CPU utilization is always higher than 100%, thereby incurring severe overload conditions. A result in this case is too many deadline misses, which cause the loop with the lowest priority, i.e. Loop 1, to become unstable, as mentioned previously.

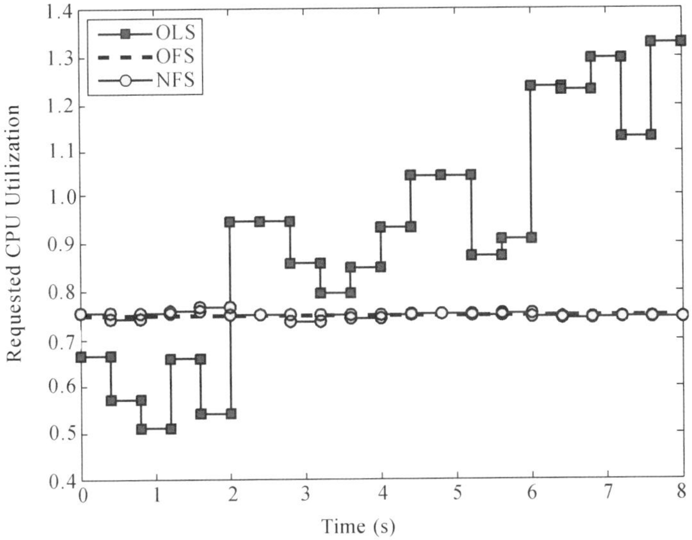

Fig. 3. 12 Total requested CPU utilization

On the contrary, OFS and NFS are able to keep the (requested) CPU utilization at or very close to the desired level $\widetilde{U}_R = 75\%$ by means of dynamic adjustment of task periods. The system schedulability is always guaranteed. Furthermore, by solving the optimization problem formulated by Eq.(3.13), OFS and NFS allocate available computing resources online in an optimal way that the total control cost is minimized under schedulability constraint.

Feedback Scheduling Overhead

Next the runtime overhead of the neural feedback scheduler is examined in comparison with the optimal feedback scheduler. Both of the feedback schedulers are implemented in the same environment using Matlab. The hardware platform is the same PC with an Intel® Celeron® CPU of 2.40 GHz. The operating system is Microsoft® Windows® XP professional edition. It should be noted that this environment cannot provide real-time guarantees. It is used here only for the comparison of different feedback schedulers in terms of time overhead. Attention should be paid to the relative values rather than the absolute values of the CPU time that feedback schedulers consume.

For the optimal feedback scheduler and the neural feedback scheduler, the CPU time they actually expend for 500 consecutive runs is recorded, respectively. In each run, task execution time is randomly drawn from the sets given in Fig.3.6.

As shown in Fig.3.13, the execution time of the optimal feedback scheduler based on the SQP method falls between 0.12 s and 0.25 s in most cases, with an average of 0.1701 s. Fig.3.14 shows that the execution time of the neural feedback scheduler is always less than 0.04 s, and close to 0.02 s in most of the cases, with an average of 0.0207 s. The ratio of the time overhead of OFS to NFS approximates 0.1701∶0.0207≈8.22∶1. That is, the overhead of NFS is only 12.2% that of OFS.

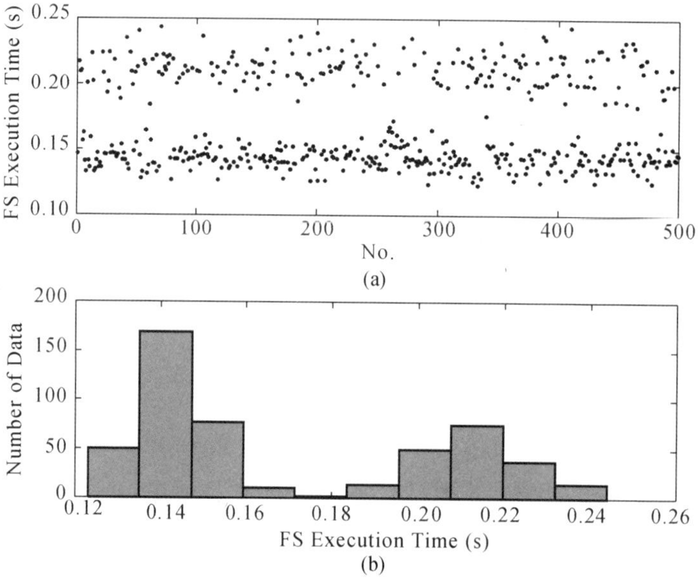

Fig. 3. 13 Execution time of optimal feedback scheduler
(a) Profile; (b) Histogram

Fig. 3.15 gives the box and whisker plot for recorded execution time of two feedback schedulers. It is clear that the neural feedback scheduler induces significantly less computational overhead than the optimal feedback scheduler. Thanks to the simple form of the cost functions, i.e. Eq.(3.11) employed in the experiments, the time overhead of the optimal feedback scheduler is not very big. Intuitively, as the form of the cost functions becomes more complex, the overhead of OFS will increase, but the overhead of NFS will not be substantially affected.

3.5 Summary

Having recognized the major disadvantages of optimal feedback scheduling, this chapter successfully incorporates the the neural networks technology into the framework of feedback scheduling. As a fast and intelligent feedback scheduling scheme, the neural feedback scheduling is presented. It fully exploits the offline solutions for the optimal feedback scheduling problem, which are offered by well-

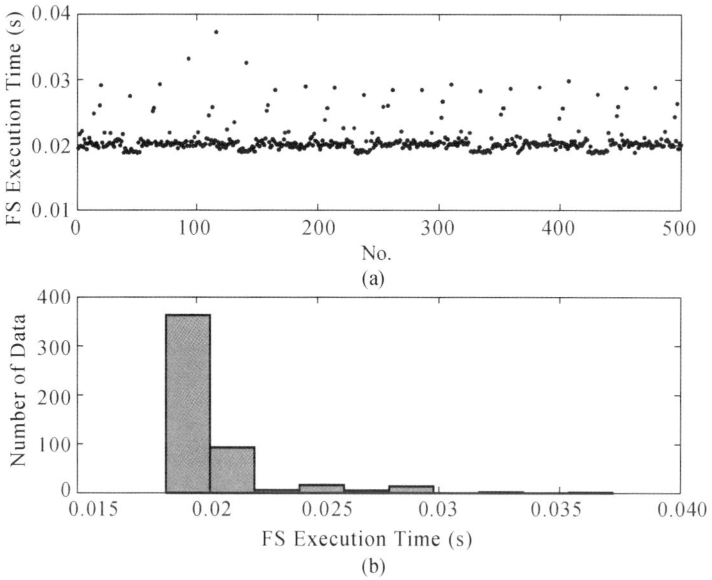

Fig. 3.14 Execution time of neural feedback scheduler
(a) Profile; (b) Histogram

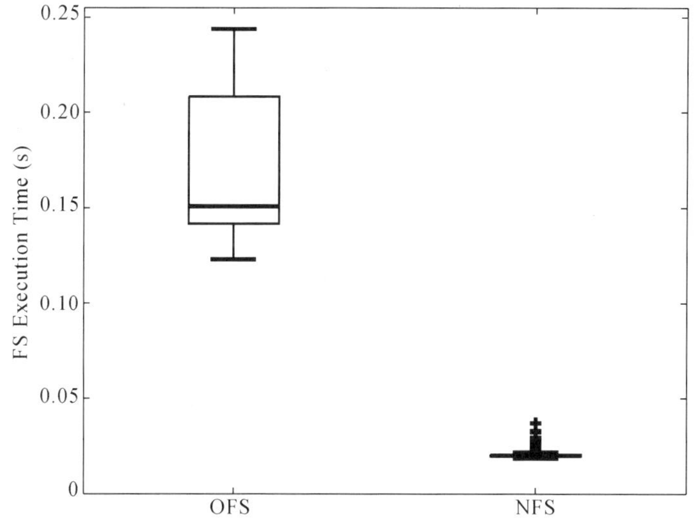

Fig. 3.15 Box and whisker plot for execution time

established mathematical optimization algorithms such as SQP. As a result, almost optimal feedback scheduling performance is achieved. The proposed approach does not rely on any specific forms of the control cost functions, making it widely applicable. Meanwhile, compared to optimal feedback scheduling, it can significantly reduce the runtime overhead, which makes it well-suited for online use. In addition, as an intelligent method, the neural feedback scheduling scheme is also characterized by good adaptability, robustness and fault-tolerance, though this is not detailed in this chapter.

Nevertheless, since the neural feedback scheduler is designed on the basis of offline solutions, neural feedback scheduling is by nature a stationary optimization method. The actual control performance (e.g. control errors) of control loops is not taken into account during runtime. Feedback scheduling schemes that exploit feedback information about the actual control performance of control loops when making decisions on resource allocation will be studied in Chapters 6 and 7.

References

[ARZ99] K.E. Årzén, B. Bernhardsson, J. Eker, A. Cervin, P. Persson, K. Nilsson, L. Sha. Integrated Control and Scheduling. *Research Report*, ISSN 0820 - 5316, Dept. Automatic Control, Lund Institute of Technology, 1999.

[ARZ03] K.E. Årzén, A. Cervin, D. Henriksson. Resource-Constrained Embedded Control Systems: Possibilities and Research Issues. *Proc. of Codesign of Embedded Real-Time Systems Workshop*, Porto, Portugal, 2003.

[AST97] K.J. Åström, B. Wittenmark. *Computer Controlled Systems: Theory and Design, 3rd Edition.* Prentice Hall, 1997.

[BOY04] S. Boyd, L. Vandenberghe. *Convex Optimization*, Cambridge Press, 2004.

[BRA02] M.S. Branicky, S.M. Philips, W. Zhang. Scheduling and Feedback Codesign for Networked Control Systems. *Proc. 41st IEEE CDC*, Las Vegas, pp.1211 - 1217, 2002.

[CAS06] R. Castañé, P. Martí, M. Velasco, A. Cervin, D. Henriksson. Resource Management for Control Tasks Based on the Transient Dynamics of Closed Loop Systems. *18th Euromicro Conf. on Real-Time Systems*, Dresden, Germany, 2006.

[CER02] A. Cervin, J. Eker, B. Bernhardsson, K.E. Årzén. Feedback-Feedforward Scheduling of Control Tasks. *Real-Time Systems*, Vol. 23, No.1, pp.25 - 53, 2002.

[CER06] A. Cervin, P. Alriksson. Optimal On-line Scheduling of Multiple Control Tasks: A Case Study. *Proc. of the 18th Euromicro Conf. on Real-Time Systems*, Dresden, Germany, 2006.

[EKE00] J. Eker, P. Hagander, K.E. Årzén. A Feedback Scheduler for Real-Time Controller Tasks. *Control Engineering Practice*, Vol. 8, No.12, pp.1369 - 1378, 2000.

[FLE87] R. Fletcher. *Practical Methods of Optimization.* John Wiley & Sons, 1987.

[HE04] J. He, H. Zhang, Y. Jing. An Integrated Control and Scheduling Optimization Method of Networked Control Systems. *Journal of Electronic Science and Technology of China*, No.02, pp.57 - 60, 2004.

[HEN05] D. Henriksson, A. Cervin. Optimal On-line Sampling Period Assignment for Real-Time Control Tasks Based on Plant State Information. *Proc. of the 44th IEEE Conf. on Decision and Control and European Control Conf. ECC*, Seville, Spain, 2005.

[JIN05] H. Jin, H. Wang, H.A. Wang, G. Dai. Optimization Design of Controller Periods Using Evolution Strategy. *CIS'05*, *Lecture Notes in Artificial Intelligence*, Vol. 3801, pp.1100 - 1105, 2005.

[KIM98] B. Kook Kim. Task Scheduling with Feedback Latency for Real-Time Control Systems. *5th Int. Conf. on Real-Time Computing Systems and Applications (RTCSA)*, Hiroshima, Japan, pp.37 - 41, 1998.

[MAR02] P. Martí. Analysis and Design of Real-Time Control Systems with Varying Control Timing Constraints. Ph.D Thesis, Technical University of Catalonia, 2002.

[SET96] D. Seto, J.P. Lehoczky, L. Sha, K.G. Shin. On Task Schedulability in Real-Time Control Systems. *Proc. 17th IEEE RTSS*, Washington, DC, pp.13 - 21, 1996.

[SU03] G. Su, F. Deng. On the Improving Backpropagation Algorithms of the Neural Networks Based on MATLAB Language: A Review. *Chinese Bulletin of Science and Technology*, Vol. 19, No.2, pp. 130 - 135, 2003.

[SUN97] Z. Sun, Z. Zhang, Z. Deng. *Theory and Technology of Intelligent Control.* Beijing: Tsinghua University Press, 1997.

[SUN04] W. Sun, C. Xu, D. Zhu. *Optimization Methods.* Beijing: Higher Education Press, 2004.

[SUN06] Y.X. Sun, J. Chu. *Industrial Process Control Technology.* Beijing: Chemical Industry Press, 2006.

[XIA05] F. Xia, Y.X. Sun. Neural Network Based Feedback Scheduling of Multitasking Control Systems. *Lecture Notes in Artificial Intelligence*, Vol. 3682, pp.193 - 199, 2005.

[XIA07] F. Xia, Y.C. Tian, Y.X. Sun, J.X. Dong. Neural Feedback Scheduling of Real-Time Control Tasks. Submitted, 2007.

[YAN04] X.H. Yang. Neural Networks with Applications to Control. Ph.D Thesis, Zhejiang University, 2004.

Chapter 4

Fuzzy Feedback Scheduling

The performance of conventional feedback scheduling schemes depends highly on the availability and accuracy of information such as task timing attributes and system workloads. It is not surprising that many practical platforms cannot provide the support that they require. To achieve satisfactory feedback scheduling performance when task execution time is unavailable and noises are present in workload measurements, an innovative feedback scheduling scheme based on fuzzy logic control is presented, namely fuzzy feedback scheduling.

The architecture for implementing fuzzy feedback scheduling is discussed, with two alternatives given. Using a simplified mobile robot target tracking system as a case study, the design methodology of the fuzzy feedback scheduler is detailed. Due to the powerful capability of fuzzy logic in dealing with nonlinearity, imprecision and uncertainty, fuzzy feedback scheduling is independent of task execution time and is robust to measurement noise, while handling uncertainties in workload and available resources in an intelligent fashion. Being easy to implement, the fuzzy feedback scheduler induces a small runtime overhead. The case study results highlight the effectiveness and merits of the proposed scheme.

4.1 Introduction

For a conventional feedback scheduling scheme, e.g. the optimal feedback scheduling described in the previous chapter, it is mandatory that the following information be available in order to calculate the accurate (optimal) sampling periods: 1) the control cost function as a function of sampling period for each control loop, 2) the execution time of control tasks, 3) the CPU utilization that determines the total allowable utilization of the control tasks. Given this information precisely, the optimal sampling periods that yield the best overall control performance can be obtained using the optimal feedback scheduling method. In practical systems, however, this may not always be the case.

In many situations, particularly in complex control systems such as nonlinear control applications, it is difficult, if not impossible, to describe the performance index of QoC as a function of the sampling period. Most often existing approximate forms of the control cost functions are applicable to only some special control systems. Their applicability to other control applications is left to be validated.

In a variety of COTS operating systems, direct measurement of task execution time is not always supported [SIM05]. In POSIX systems, for example, it is often not easy to measure online the execution time of a task, though other timing attributes such as response time may be obtained quite simply.

Additionally, measurement noises of certain magnitudes inevitably exist in real-world applications, even though online measurement of task execution time is supported by the underlying operating system kernel. In a like manner, available measurements or estimations of the system workload are also imprecise due to noises. Furthermore, there are many other factors such as the invocation interval of the feedback scheduler that also affect the accuracy of measurements of the CPU utilization [AMI05]. When the availability and accuracy of critical information cannot be guaranteed, conventional feedback scheduling schemes may not perform as expected, even becoming unfeasible in some circumstances.

This chapter considers real-world embedded control systems with the following properties: 1) the online availability of information such as task execution time cannot be guaranteed, 2) system parameter measurements are imprecise. To perform efficient feedback scheduling when accurate timing parameters are not available, this chapter takes advantage of feedback control, and presents a Fuzzy Feedback Scheduling (FFS) scheme [XIA05a, XIA05b, XIA07]. Integrating the fuzzy logic control technique [KEV98, SUN97, SUN06] with the feedback scheduling framework, this scheme has the following potential advantages:

(1) Fuzzy control builds controllers based on heuristic information that comes from a practitioner acting as a human-in-the-loop controller for a particular process. Modelling of the process to be controlled is not required for fuzzy control system design. As a consequence, feedback scheduling based on fuzzy logic control does not depend on explicit formulation of the relationship between control cost functions and sampling periods.

(2) As a formal methodology to emulate the intelligent decision-making process of a human expert, fuzzy control provides a simple and flexible way to arrive at a definite conclusion based on imprecise, noisy, or incomplete input information. Therefore, it can serve as an enabling technology for feedback scheduling in the absence of task execution time as well as in the presence of measurement noises.

(3) Fuzzy controllers are simple in structure, and are easy to implement and use. The small amount of computation makes it well-suited for satisfying the low overhead requirement of feedback scheduling.

(4) Fuzzy control is robust and adaptable in that it can deliver good performance regardless of whether or not the controlled process is linear. These capabilities will reinforce good performance of feedback scheduling in dynamic and unpredictable environments.

Fuzzy feedback scheduling establishes a mapping from feedback scheduling to fuzzy logic control. With the goal of improving the overall control performance as much as possible, the fuzzy feedback scheduler attempts to keep the actual CPU utilization at a desired level in an intelligent manner. In this way, intelligent and flexible QoC management is facilitated.

Two alternative architectures are suggested for implementing fuzzy feedback scheduling. The basic flow of the feedback scheduling algorithm employed in this chapter is described. The design of a fuzzy feedback scheduler is illustrated using a simplified mobile robot target tracking system as a case study. Some critical design issues are discussed. In contrast to conventional feedback scheduling schemes, the proposed fuzzy feedback scheduling is independent of task execution time, and capable of handling the uncertainty, imprecision and incompletion of system parameters. Simulation results of the case study show that the fuzzy feedback scheduling scheme performs quite well in the presence of workload uncertainties and measurement noises.

In recent years, applying fuzzy control techniques to general-purpose computing and communication systems has attracted increasing attention from both academic and engineering sides, for example, the control of server systems [DIA02] and network congestion control [REN02]. There are also some attempts to introduce this technique into the area of real-time task scheduling. For instance, Zou [ZOU03] uses fuzzy control technology to design an adaptive scheduling method for soft real-time tasks in open real-time systems, where the rate of the server for each task is adapted with respect to workload variations. Jin, *et al.* [JIN04] explore a fuzzy scheduling algorithm within the framework of feedback scheduling and integrate the methodology of fuzzy control. The goal of the work is to provide QoS guarantees for real-time scheduling with uncertain task timing attributes and workload. However, none of this work deals with control applications.

Lei, *et al.* [LEI05] employ fuzzy control in constructing feedback schedulers for Real-Time Control Systems. They develop a fuzzy controller to dynamically adjust the priorities of control tasks according to control errors and their change rates. The feedback scheduler allocates CPU utilization among tasks with respect to newly assigned priorities. Although fuzzy control technology is used therein, it is mainly for modifying task priorities, which is completely different from the function of the fuzzy feedback scheduler developed in this chapter. Moreover, Lei, et al. [LEI05] actually distributes available computing resources among tasks based on the knowledge of task priorities. On the contrary, this chapter adapts sampling periods in a different way that the CPU utilization is controlled by means of fuzzy control, thus holistically integrating fuzzy control with feedback scheduling.

The remainder of this chapter is structured as follows. Section 4.2 describes the problem considered, with the basic idea of fuzzy feedback scheduling illustrated from the control perspective. Section 4.3 gives two alternative architectures for implementing fuzzy feedback scheduling. After discussing the advantages and disadvantages qualitatively, the basic structure of the feedback scheduling algorithm is described. In Section 4.4, the design procedures for the fuzzy feedback scheduler are detailed, in the context of a case study on a simplified mobile robot system. Some critical design issues are also discussed. Section 4.5 assesses the performance of fuzzy feedback scheduling via simulation experiments, with comparison against the traditional open-loop scheduling scheme and an ideal feedback scheduling scheme. Finally, Section 4.6 summarizes this chapter.

4.2 Problem Description

Consider a processor with limited processing power, upon which N independent control tasks run concurrently. Each control task executes a well-designed control algorithm that is responsible for the control of a physical process. Besides these control tasks, non-control tasks for data backup, human machine interaction, and remote communication may possibly exist. For simplicity, all tasks in the system are assumed to be periodic.

The major timing attributes of control task i are defined below:

(1) h_i: period, which is equal to the sampling period of corresponding control loop, and is available precisely online.

(2) c_i: execution time, which is time-varying and unavailable.

In systems where measuring task execution time is not supported, most often it is possible to obtain the estimation of CPU workload (or total utilization). For example, Simon, *et al.* [SIM05] estimate the CPU workload in a robot control system through recording response time of tasks. Another possible approach to estimate CPU workload is to divide the total amount of system busy time within every feedback scheduler invocation interval T_{FS} by T_{FS}. Certainly, measurement noises are unavoidable no matter what method is used to measure or estimate the runtime CPU utilization. Without loss of generality, assume the measurement of CPU utilization is:

$$\hat{U} = \sum_{i=1}^{N} \frac{c_i}{h_i} + U_{\text{others}} + \delta_U, \tag{4.1}$$

where U_{others} represents the total CPU utilization of all non-control tasks, and δ_U is measurement noise. Notice that the above equation will be used only for calculating the feedback of the CPU utilization in simulation experiments. The design of fuzzy feedback schedulers does not rely on any specific types of formulas for computing the CPU utilization measurement. In the system considered here, c_i and U_{others} are unavailable for the feedback scheduler. The information that the feedback scheduler can collect includes the utilization measurement $\hat{U}$ and sampling periods h_i.

To achieve flexible QoC management, the problem of feedback scheduling of the above system is examined from a viewpoint of feedback control. The general framework of feedback scheduling given in Fig.2.10 is employed. Since measuring the CPU utilization of each task is not supported, this chapter attempts to maintain the total CPU utilization of all tasks (including both control tasks and non-control tasks) at a desired level. Accordingly, the controlled variable associated with the feedback scheduling loop is the CPU utilization. In the context of feedback scheduling, the most common choice of the manipulated variable is task periods, which are equal to the sampling periods of the control loops. Thus the feedback scheduler will dynamically adapt the sampling periods of each control loop to control the CPU utilization. It is noteworthy that the timing attributes of non-control tasks cannot be changed intentionally by any feedback schedulers.

Let U_R denote the desired CPU utilization. It is intuitive that U_R should not violate the schedulability constraint of the system. For instance, if EDF is employed as the underlying scheduling policy, it should then hold that $U_R \leqslant 1$. Naturally, some schedulability margin must be preserved when choosing U_R in practice, i.e. $\varepsilon = 1 - U_R > 0$, because of the presence of measurement noises.

According to the above description, the relevant feedback scheduling problem can be stated as follows: *taking into account time-varying, unavailable task execution time and imprecise CPU utilization measurements in a multitasking embedded control system, design an efficient feedback scheduler to achieve flexible QoS management based on utilization control*. Due to the unavailability of task execution time, conventional feedback scheduling strategies that depend on complete formulation of task models become unfeasible. Furthermore, the imprecision of temporal parameters is a practical issue that conventional feedback schedulers cannot cope with, which may have significant (negative) impact on the feedback scheduling performance.

To address the above problem, this chapter suggests an intelligent feedback scheduling approach based on fuzzy control. In the following, the architecture and design methodology of this approach will be detailed.

4.3 Framework

To implement fuzzy feedback scheduling, an appropriate architecture should be specified. This section first discusses two alternative kinds of system architectures for realizing fuzzy feedback scheduling, and then describes the basic flow of the algorithm used.

4.3.1 Alternative System Architectures

There are two alternative methods for realizing feedback scheduling systems. The first is that one separate feedback scheduler is designed for each control loop, whose responsibility is adjusting the sampling periods of the corresponding control loop according to the measured total CPU utilization. The second is to design a global feedback scheduler that adjusts simultaneously the sampling periods of all control loops. To distinguish between these two methods, they are called *decentralized feedback scheduling* and *centralized feedback scheduling*, respectively.

Decentralized Feedback Scheduling

As mentioned above, the most distinct feature of this method is that every control loop has its own feedback scheduler to determine the sampling periods. The architecture of this kind of feedback scheduling systems is shown in Fig. 4.1. It is clear that there are a total of N feedback schedulers in this case.

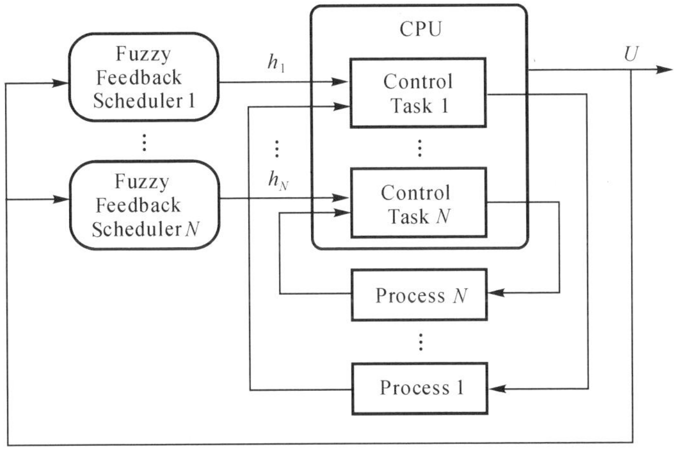

Fig. 4.1 Structure of decentralized feedback scheduling

Since each task period is adjusted by a separate feedback scheduler, it is quite flexible to implement feedback scheduling by exploiting this architecture. The different requirements of diverse control loops can be met simultaneously by multiple feedback scheduling loops. Nevertheless, the design of the feedback schedulers could be quite hard. Some of the reasons behind are given below.

First, it is obligatory to design a specific feedback scheduler for each control loop, which complicates the work at design time, and also increases the feedback scheduling overhead at runtime. Second, when designing a feedback scheduler for a certain control loop, one must take into account the constraint on the allowable CPU utilization that derives from all other tasks so as to preserve system schedulability. This makes feedback scheduler design even harder. Moreover, additional coordination mechanisms may be required for running multiple feedback schedulers in parallel.

Centralized Feedback Scheduling

In contrast to decentralized feedback scheduling, centralized feedback scheduling employs one single global feedback scheduler to adapt all sampling periods simultaneously, as shown in Fig.4.2. This implies that only one feedback scheduler needs to be designed in this case. The system complexity is reduced to some extent as compared with the de-centralized approach. It is noteworthy that in the context of fuzzy feedback scheduling, the feedback scheduler in general becomes a typical multi-input multi-output fuzzy controller. It is often difficult to design directly such a fuzzy feed-back scheduler.

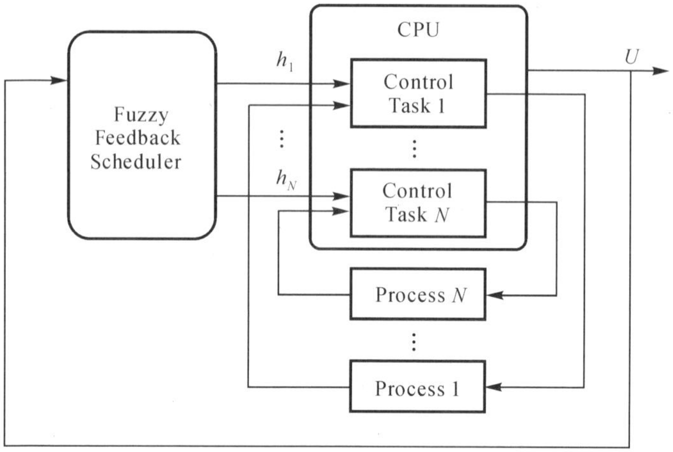

Fig. 4. 2 Structure of centralized feedback scheduling

4.3.2 Basic Flow of the Algorithm

This chapter employs the second system architecture (see Fig.4.2) for realizing fuzzy feedback scheduling. The feedback scheduler is time-triggered. To reduce the difficulty in feedback scheduler design, a simple rescaling method given below is used to determine the sampling periods:

$$h_i(j) = \eta(j) h_i(j-1), \ i = 1, \cdots, N, \tag{4.2}$$

where $\eta(j)$ is the *period rescaling factor*, and j denotes the invocation instant of the feedback scheduler. The above rescaling method makes control tasks behave like the elastic scheduling model [BUT02]. In [CER02] Cervin, *et al.* also exploit this method delivering quite good performance.

Taking into account the constraint of the maximum allowable sampling period $h_{i,\max}$, the calculation of sampling periods is expressed as:

$$h_i(j) = \min\{h_{i,\max}, \ \eta(j) h_i(j-1)\}, \ i = 1, \cdots, N. \tag{4.3}$$

Ideally, if the timing attributes of all tasks are precisely known, then the period rescaling factor can be set to:

$$\eta(j) = \frac{\sum_{i=1}^{N} \frac{c_i(j)}{h_i(j-1)}}{U_R - U_{\text{others}}(j)}. \tag{4.4}$$

After the sampling periods are altered using Eq.(4.3) along with Eq.(4.4), the total CPU utilization of control tasks in the next invocation interval will be:

$$\begin{aligned} U(j) &= \sum_{i=1}^{N} \frac{c_i(j)}{h_i(j)} + U_{\text{others}}(j) = \frac{1}{\eta(j)} \sum_{i=1}^{N} \frac{c_i(j)}{h_i(j-1)} + U_{\text{others}}(j) \\ &= \frac{U_R - U_{\text{others}}(j)}{\sum_{i=1}^{N} \frac{c_i(j)}{h_i(j-1)}} \cdot \sum_{i=1}^{N} \frac{c_i(j)}{h_i(j-1)} + U_{\text{others}}(j) = U_R \end{aligned} \quad . \tag{4.5}$$

Obviously, using this ideal method one will achieve the desired CPU utilization in the next invocation interval. However, this is only true for the ideal case where all necessary system parameters are precisely known. In real systems described in Section 4.2, it is impossible to use Eq.(4.4) to compute the period rescaling factor.

In this chapter the fuzzy control technique is used to determine the period rescaling factor η. Hence, the fuzzy controller to be designed has only one output, i.e. η. As a consequence, the design and implementation of the fuzzy feedback scheduling system shown in Fig.4.2 becomes quite simple. That is why Eq.(4.3) is used to rescale all sampling periods with the same factor.

The work flow of the fuzzy feedback scheduling system can be outlined as follows. At every invocation instant, the feedback scheduler samples the CPU utilization that the system has been monitoring from the previous invocation instant, and compares it with its desired value. Based on the control error of CPU utilization and the change in error, the fuzzy controller acting as the feedback scheduler produces the corresponding period rescaling factor. The sampling period of each control loop is then re-assigned using Eq.(4.3).

4.4 Fuzzy Feedback Scheduler Design

Now that the architecture and the work flow of the algorithm have been determined, the design of the fuzzy feedback scheduler becomes the key to fuzzy feedback scheduling. Since the fuzzy feedback scheduler is a fuzzy controller from the control perspective, this chapter uses these terms interchangeably when referring to the feedback scheduler. Based on the framework given in the previous section, this section will detail the design procedures for the fuzzy feedback scheduler and discuss some critical design issues. For this purpose, a simplified mobile robot target tracking system is used for the case study.

4.4.1 A Case Study

Consider a simplified mobile robot system [AND05] as shown in Fig.4.3. The robot is regarded as a point (x, y) on the plane. It can move on x-axis and y-axis freely and independently. The coordinate of the robot on each axis, i.e. x and y, is controlled by a separate control loop respectively. For example, the position may be controlled through manipulating the current to the relevant motor. The overall control goal of the mobile robot system is to track as closely as possible a mobile target, which is also modelled as a (mobile) point on the plane.

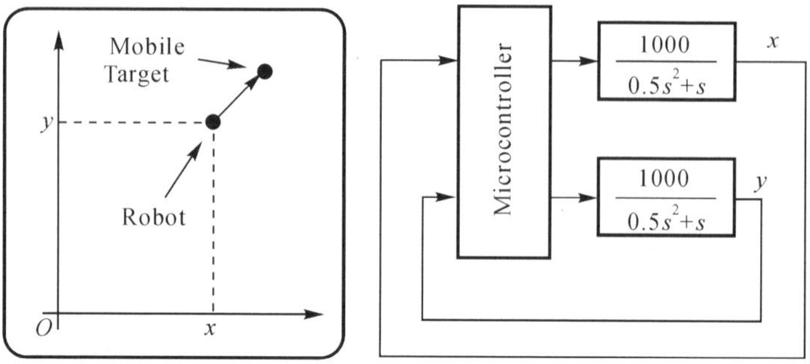

(a) (b)

Fig. 4.3 Simplified mobile robot system
(a) Coordinates of Mobile Robot; (b) Control of Coordinates

The transfer functions of two control loops are the same as $G(s) = 1000/(0.5s^2 + s)$. Controllers are designed by discretization of continuous-time controllers that use the PID algorithm, see Chapter 6 for details on the forms. The controller parameters are set in accordance with [MAR02]. Two control tasks (denoted τ_1 and τ_2) share the processor with a third periodic non-control task τ_3, which may take the responsibility of higher-level decision-making, coordination and communication with other devices.

4.4.2 Design Methodology

Fig.4.4 gives the block diagram for the internal structure of the fuzzy feedback scheduler, where the component of sampling period adjustment is omitted for the sake of simplicity. Like almost all fuzzy controllers for control applications, the fuzzy feedback scheduler mainly consists of four components, i.e. a fuzzification interface, a rule-base, an inference mechanism, and a defuzzification interface. Inputs to the fuzzy feedback scheduler are the CPU utilization control error $e(j) = U_R - \hat{U}(j)$ and the change in error $ec(j) = e(j) - e(j-1)$, while the output is naturally the sampling period rescaling factor η.

A fuzzy system represents a nonlinear mapping between its inputs and outputs, which are real numbers, but not fuzzy sets. The inner work flow of the fuzzy feedback scheduler is as follows. First, the fuzzification interface translates numeric inputs $e(j)$ and $ec(j)$ into fuzzy sets characterizing linguistic variables E and EC. The inference mechanism then activates and applies a predetermined set of linguistic rules in the rule-base with respect to these linguistic variables, and produces the fuzzy sets of the output linguistic variable RF. Finally the defuzzification interface converts the fuzzy conclusions when the inference mechanism reaches to a numeric value $\eta(j)$ that can be used in Eq.(4.3).

To simplify online computations, the look-up table method [SUN97, SUN06, WEN01, ZOU03] is used to establish the fuzzy feedback scheduler. As shown in Fig.4.4(b), the rational of this method is as follows. A fuzzy control table (also

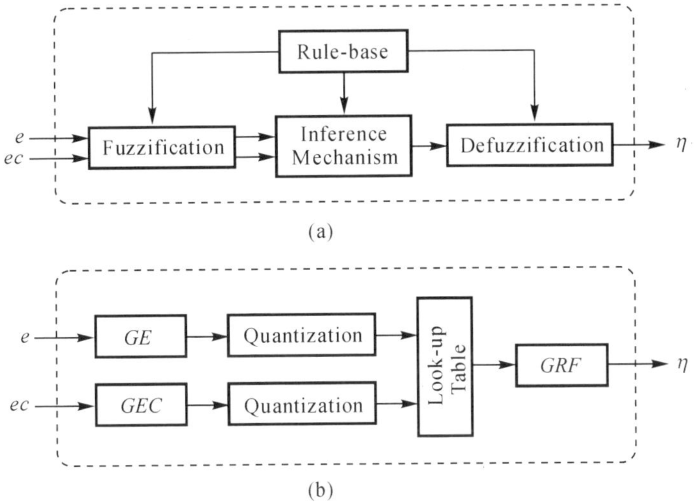

Fig. 4.4 Internal structure of fuzzy feedback scheduler
(a) Offline Computation; (b) Online Computation

called look-up table) is first constructed via offline computations using the original fuzzy feedback scheduler depicted in Fig.4.4(a), and store it in the memory. During runtime, the resulting feedback scheduler quantizes the inputs $e(j)$ and $ec(j)$ using the input scaling factors GE and GEC, and then searches in the fuzzy control table to find out the quantized output value corresponding to the quantized inputs. The final output $\eta(j)$ is produced by multiplying the quantized output value with the output scaling factor GRF.

The key to this method is building the fuzzy control table, on which the feedback scheduling performance depends heavily. The design procedures of the fuzzy feedback scheduler are detailed below, while using the mobile robot system given in Fig.4.3 as the target application.

Specify the structure of fuzzy feedback scheduler

According to the above description, a two-dimension fuzzy controller is used, with two input variables (e and ec) and one output η. The universes of discourse for e, ec, and η are chosen to be $[-0.3, 0.3]$, $[-0.3, 0.3]$, and $[0.5, 1.5]$, respectively.

Describe linguistically inputs and outputs

In designing the fuzzy feedback scheduler, we employ the following settings for inputs and outputs. Both of the sets of linguistic values for the linguistic variables E and EC are

$$\{\text{NB, NS, ZE, PS, PB}\},$$

and the set of linguistic values for RF is

$$\{\text{NB, NM, NS, ZE, PS, PM, PB}\},$$

where NB represents *negative big*; NM represents *negative medium*; NS represents *negative small*; ZE represents *zero*; PS represents *positive small*; PM represents *positive medium*; PB represents *positive big*.

To facilitate the construction of the look-up table, both inputs and outputs should be quantized. Both of the sets of quantized values for the inputs are

$$\{-6, -5, -4, -3, -2, -1, 0, 1, 2, 3, 4, 5, 6\}.$$

The set of quantized values for the outputs is

$$\{-7, -6, -5, -4, -3, -2, -1, 0, 1, 2, 3, 4, 5, 6, 7\}.$$

Accordingly, $GE = 6/0.3 = 20$, $GEC = 6/0.3 = 20$, $GRF = 1/14 = 0.0714$.

Illustrated below is how to quantize inputs and how to de-quantize outputs. Suppose that e or ec is equal to a real number α. The corresponding quantized value β is calculated as

$$\beta = \left[GE \cdot \left(\alpha - \frac{0.3 - 0.3}{2}\right)\right] = [20\alpha], \tag{4.6}$$

where the operator $[\cdot]$ obtains the nearest integer around a real number.

Similarly, if the quantized value of the output η is χ, then the corresponding sampling period rescaling factor ζ will be

$$\zeta = GRF \cdot \chi + \frac{0.5 + 1.5}{2} = 0.0714\chi + 1. \tag{4.7}$$

Specify membership functions for linguistic values

Fig.4.5 depicts the membership functions for all linguistic values for both input and output linguistic variables. Accordingly, the membership values for each input and output are summarized in Tables 4.1 and 4.2, respectively.

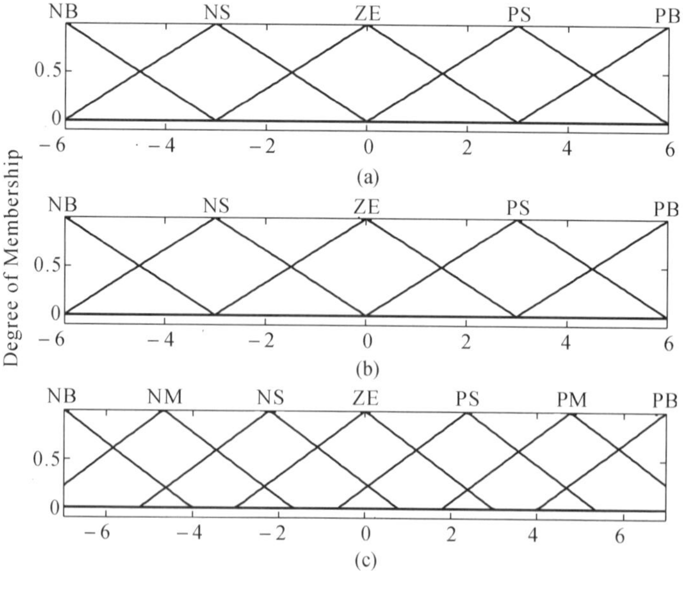

Fig. 4.5 Input and output membership functions
(a) E; (b) EC; (c) RF

Table 4.1 Membership values for linguistic input variables E and EC

	E or EC												
	-6	-5	-4	-3	-2	-1	0	1	2	3	4	5	6
PB	0	0	0	0	0	0	0	0	0	0	0.3	0.7	1.0
PS	0	0	0	0	0	0	0	0.3	0.7	1.0	0.7	0.3	0
ZE	0	0	0	0	0.3	0.7	1.0	0.7	0.3	0	0	0	0
NS	0	0.3	0.7	1.0	0.7	0.3	0	0	0	0	0	0	0
NB	1.0	0.7	0.3	0	0	0	0	0	0	0	0	0	0

Table 4.2 Membership values for linguistic output variable RF

	RF														
	-7	-6	-5	-4	-3	-2	-1	0	1	2	3	4	5	6	7
PB	0	0	0	0	0	0	0	0	0	0	0	0	0.3	0.6	1.0
PM	0	0	0	0	0	0	0	0	0	0.1	0.4	0.7	0.9	0.6	0.3
PS	0	0	0	0	0	0	0	0.2	0.5	0.9	0.8	0.5	0.1	0	0
ZE	0	0	0	0	0	0.3	0.7	1.0	0.7	0.3	0	0	0	0	0
NS	0	0	0.1	0.4	0.7	0.9	0.6	0.3	0	0	0	0	0	0	0
NM	0.2	0.6	0.9	0.8	0.4	0.1	0	0	0	0	0	0	0	0	0
NB	1.0	0.6	0.3	0	0	0	0	0	0	0	0	0	0	0	0

Determine fuzzy control rules and inference mechanism

The linguistic rules are established on the basis of analysis of the system behaviour under different circumstances, as well as related control theory and real-time scheduling theory. As shown in Table 4.3, totally 25 linguistic rules are built in this chapter for the mobile robot system.

Table 4.3 Linguistic rules

RF		EC				
		NB	NS	ZE	PS	PB
E	NB	PB	PB	PB	PB	PM
	NS	PB	PB	PM	PS	ZE
	ZE	PM	PS	ZE	ZE	NS
	PS	PS	ZE	ZE	NS	NM
	PB	ZE	NS	NM	NB	NB

The basic rational for creating the fuzzy control rules is as follows. If the measured CPU utilization exceeds the desired level significantly, a big sampling period rescaling factor should be used so that the sampling periods are enlarged quickly enough to avoid too many deadline misses. If the measured CPU utilization is lower than the desired level, the sampling periods should be reduced slowly to reduce the possibility of overload, which would also give rise to deadline misses.

To describe how to determine the linguistic rules, consider for example the case where e is negative big, implying that the actual CPU utilization is far bigger than the desired level. In this context, the requested CPU utilization should be reduced quickly no matter what the value of ec is. Therefore, η should be set to the maximum, i.e. positive big, if ec is slightly negative, which means that the requested utilization still has the tendency to increase. If ec is positive big implying that the requested CPU utilization has the tendency of significant decrease, the value of η is set to positive medium in order to avoid serious fluctuations within the feedback scheduling system. Under all other circumstances, η is chosen to be positive big to minimize the possibility of system overload. Accordingly, the following rules are obtained:

IF e is NB and ec is NB, THEN η is PB;
IF e is NB and ec is NS, THEN η is PB;
IF e is NB and ec is ZE, THEN η is PB;
IF e is NB and ec is PS, THEN η is PB;
IF e is NB and ec is PB, THEN η is PM.

In similar way, all other rules given in Table 4.3 are established.

For the inference mechanism, the max-min method is adopted. In the defuzzification interface, the most popular *centre of gravity* method is used to produce a real number in the universe of discourse of the output. In other words, the popular Mamdani inference method is used.

Create look-uptable

The last step of offline design is generating the fuzzy control table. Previous design procedures accomplish the offline computation shown in Fig. 4.4(a). To generate a look-up table that can be used online, as shown in Table 4.4, the quantized value of the output corresponding to each pair of quantized values of the input variables is calculated.

The input-output surface of the fuzzy inference system that corresponds to Table 4.4 is depicted in Fig. 4.6, which describes more straightforwardly the mapping between the input to the output conceived by the look-up table. It is easy to find that the period rescaling factor decreases as the error in utilization control increases. Particularly, if $e < 0$, η will be larger than its average, i.e. $\eta > (0.5 + 1.5)/2 = 1$; otherwise, if $e > 0$, η will then be smaller than its average, i.e. $\eta < 1$.

Once the look-up table is constructed, it can be used during runtime within the framework given in Fig.4.4(b). The work flow of the fuzzy feedback scheduling algorithm is illustrated in Fig.4.7.

Table 4.4 Look-up table for fuzzy feedback scheduler

η		ec												
		-6	-5	-4	-3	-2	-1	0	1	2	3	4	5	6
	-6	6	6	6	6	6	6	6	6	6	6	5	5	5
	-5	6	6	6	6	5	5	5	4	4	4	3	3	3
	-4	6	6	6	6	5	5	5	4	3	3	2	2	2
	-3	6	6	6	6	5	5	5	4	3	2	2	1	0
	-2	5	4	4	4	3	3	3	3	2	2	1	0	-1
	-1	5	4	3	3	2	2	2	2	2	1	0	0	-1
e	0	5	4	3	2	2	1	0	0	0	0	-1	-1	-2
	1	4	3	2	2	2	1	0	-1	-1	-1	-2	-2	-3
	2	3	2	2	1	1	1	0	-1	-1	-1	-2	-3	-4
	3	2	2	1	0	0	0	0	-1	-1	-2	-3	-4	-5
	4	2	1	0	-1	-2	-2	-2	-2	-2	-3	-3	-4	-5
	5	1	0	0	-1	-2	-3	-3	-3	-3	-4	-4	-4	-5
	6	0	-1	-1	-2	-3	-4	-5	-5	-5	-6	-6	-6	-6

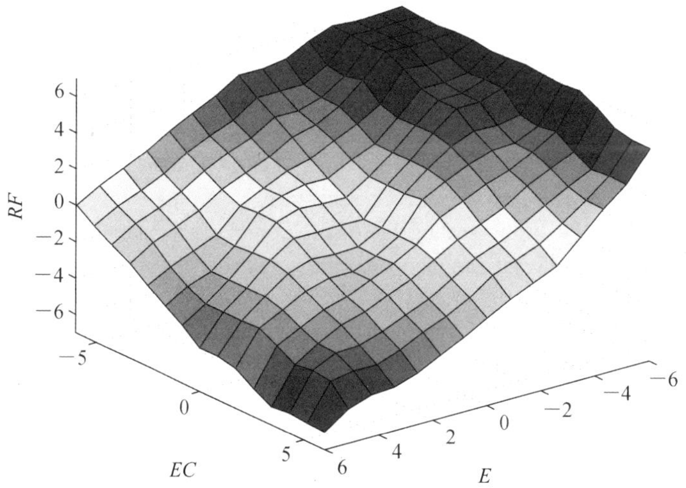

Fig. 4.6 Input-output surface for fuzzy feedback scheduler

The program of the fuzzy feedback scheduling algorithm is fairly simple. With only a small amount of computational effort, it is easy to implement. The feedback scheduler is able to produce the period rescaling factor through limited computations and searching on a table of 13×13 dimensions. With a time complexity of $O(1)$, the computational time of the feedback scheduling algorithm is insignificant, which leads to quite a small feedback scheduling overhead.

4.4.3 Design Considerations

Just like the design of a general fuzzy controller, care must be taken in designing the

```
//Û: Measured CPU utilization
//h: Sampling period
Fuzzy Feedback Scheduling {
    Input: Û
    //e_old: control error at last instant
    //GE, GEC: input scaling factors
    //GRF: output scaling factor
    Compute e ← U_R−Û;
    Compute ec ← e−e_old;
    Compute E ← [GE×e];
    Compute EC ← [GEC×ec];
    Seach RF in the look−up table;
    Compute η ← GRF×RF+1;
    Compute h using Eq.(4.3);
    //Update internal variables
    e_old ← e;
    Output: h
}
```

Fig. 4.7 Pseudo code for fuzzy feedback scheduling algorithm

fuzzy feedback scheduler when making decisions about design parameters such as input and output scaling factors, quantization levels, and invocation interval. These critical issues are discussed below based on general knowledge of fuzzy control theory [SUN97, SUN06]. Some instructions are given concerning how to make better the design of the fuzzy feedback scheduler.

The choice of the input scaling factors, *GE* and *GEC*, and the output scaling factor *GRF* has significant impact on the performance of the fuzzy feedback scheduler. In principle, if *GE* increases, which is equivalent to shrinking the universe of discourse of the CPU utilization control error, then the effect of the error will be enhanced. As a consequence, the rise time of the feedback scheduling system will be shortened. Due to larger overshoots, however, the transient process of the feedback scheduling system will become longer. Decreasing *GE* reduces the magnitude of overshoot, but also decreases the steady-state accuracy for controlling the CPU utilization.

It is possible to prevent the system from suffering from overshoot through properly choosing the value of *GEC*. As *GEC* increases, the magnitude of overshoot decreases accordingly. The major penalty is that the response of the system becomes slower. In a sense, different values of *GE* and *GEC* represent different weights on the input variables, i.e. error and error change, respectively. Furthermore, they influence each other.

Not surprisingly, the value of *GRF* also impacts the behaviour of the fuzzy feedback scheduler. An overly small *GRF* will lengthen the transient process of the system, while an unnecessarily large *GRF* will possibly cause the system to oscillate. In practical applications there usually exists more than one feasible choice for the input and output scaling factors.

The choice of the quantization levels for the universes of discourse also influences the performance of the feedback scheduling loop. When determining the quantization levels, tradeoffs have to be made between accurate approximation and low complexity. It is generally believed that increasing the number of quantization levels could improve the control accuracy of the feedback scheduling system. However, this is subject to the constraint on the computer string length, and more quantization levels increase the dimension of the look-up table, which results in larger feedback scheduling overheads.

Generally speaking, the quantization of the universes of discourse can be either symmetrical or unsymmetrical. In most cases the symmetrical quantization will be used. However, there are also some situations in which the unsymmetrical quantization is preferable.

It is a common issue to choose an appropriate invocation interval when designing time-triggered feedback schedulers. The impact of the invocation interval T_{FS} is discussed here from a viewpoint of fuzzy control. In the fuzzy feedback scheduler, the two input variables, in this case the CPU utilization control error and the error change relate to two consecutive invocation intervals. In order to achieve effective control of the CPU utilization, the change in error should be kept large enough. For this reason, T_{FS} should not be too small. On the other hand, however, a smaller value of T_{FS} is preferable to deliver fine-grained control. Most often tradeoffs must be made between these conflicting requirements. A practical method is to conduct simulations for different invocation intervals, and choose the best one.

4.5 Simulation Experiments

This section conducts simulation experiments on the case study system described in Section 4.4.1 to assess the performance of the fuzzy feedback scheduler designed above.

4.5.1 Setup Overview

Suppose the target that the mobile robot tracks moves along a half circle, with a constant angular speed. It starts moving from point (0, 0) at time $t = 0$, and reaches point (2, 0) at time $t = 4$ s. The default unit on the plane is meter (m). Accordingly, the reference inputs of control loops for x and y coordinates are depicted in Fig. 4.8.

During runtime the actual execution time of three tasks is generated by $c_i = (1 + \varepsilon)\bar{c}_i$, where ε is a sequence of white Gaussian noise with zero mean and a variance of 0.01, and $\bar{c}_i$ is the average execution time of each task, given in Fig. 4.9 Notice that all task execution time is unavailable to the feedback scheduler.

The nominal period for each control task is 3, 4 and 5 ms, respectively, where the period of non-control Task 3 (i. e. h_3) is unchangeable for the feedback scheduler. Both of the maximum allowable sampling periods of two control loops

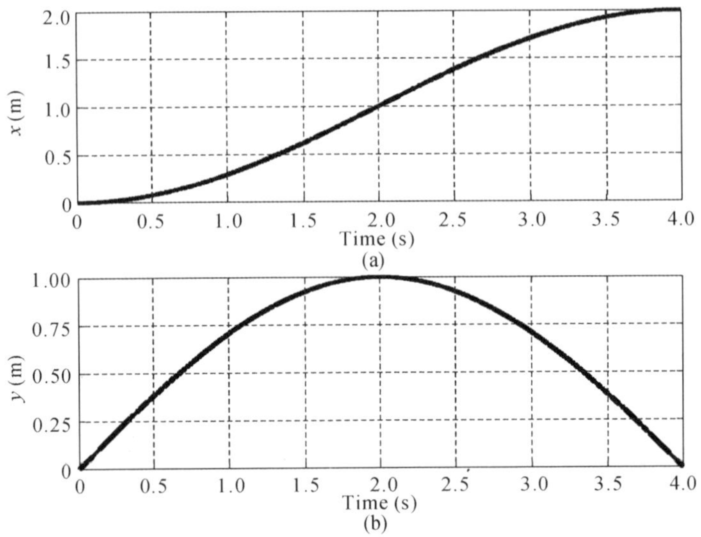

Fig. 4.8 Reference inputs for tracking control of mobile robot
(a) x-axis; (b) y-axis;

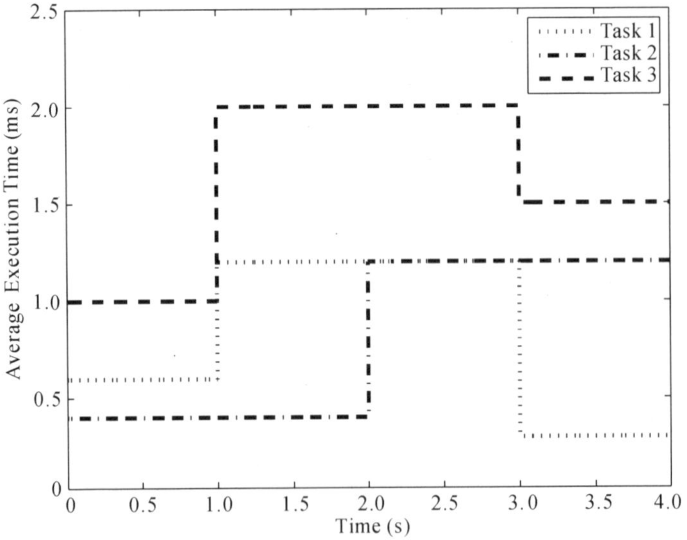

Fig. 4.9 Average execution times of tasks

are $h_{\max} = 7$ ms.

Fixed priorities are utilized in the system. Implemented as a periodic task, the feedback scheduler has the highest priority. The second highest priority is assigned to the non-control task τ_3, while τ_2 has the lowest priority. In the feedback scheduling loop the desired CPU utilization is set to $U_R = 85\%$. The invocation interval is $T_{FS} = 20$ ms, and the execution time of the feedback scheduler task is assumed to be 0.1 ms. The measured CPU utilization is generated by $\hat{U} = c_1/h_1 + c_2/h_2 + c_3/h_3 + \delta_U$, where $0 \leqslant \hat{U} \leqslant 1$, and δ_U is a sequence of zero-mean white Gaussian noise with a variance of r^2.

In this section the proposed fuzzy feedback scheduling method is compared with the following two methods:

(1) Traditional Open-Loop Scheduling (OLS): No feedback scheduling strategy is used, and control loops always run with their nominal sampling periods;

(2) Ideal Feedback Scheduling (Ideal-FS): Actual CPU utilization and all tasks' timing parameters are precisely known, and the feedback scheduler adapts sampling periods using Eqs.(4.3) and (4.4).

As a performance metric for tracking control of the mobile robot, the tracking error is defined as the distance between the robot and the target:

$$ERR_{\text{track}} = \sqrt{(x_{\text{act}} - x_{\text{ref}})^2 + (y_{\text{act}} - y_{\text{ref}})^2}, \tag{4.8}$$

where the tuple $(x_{\text{act}}, y_{\text{act}})$ denotes the current coordinates of the robot, and the tuple $(x_{\text{ref}}, y_{\text{ref}})$ denotes the coordinates of the mobile target.

4.5.2 Results and Analysis

Under open-loop scheduling, the track of the mobile robot and the tracking error are given in Fig.4.10, where solid line shows the track of the target and the centre of circles corresponds to the position of the robot. There is no robot track on the right part of Fig.4.10(a) because the robot is so far from the target that its position is out of the scope of the figure.

After time $t = 2$ s, the robot becomes unable to track the mobile target effectively. Finally the system goes unstable. By examining the average system workload one can easily find out the reason behind the system instability. In the time interval from $t = 2$ s to 3 s, the average workload of the system, which is defined as the sum of all tasks' average requested CPU utilization, is $(1.2/3 + 1.2/4 + 2/5) \times 100\% = 110\%$. It is apparent that the system is overloaded, and therefore the task set is unschedulable. Furthermore, since Task 2 has the lowest priority, the control loop for the y coordinate of the robot encounters severe deadline miss. This prevents the robot from well tracking the target.

Fig.4.11 gives the close-up of CPU schedule around the time instant $t = 2$ s when the open-loop scheduling scheme is employed. The execution of the non-control task τ_3 is not affected by control tasks because of its higher priority; on the contrary, control task τ_2 is preempted frequently because its priority is the lowest, which explains why this loop finally becomes unstable as well as why the track of the mobile robot is out of control.

The performance of the ideal feedback scheduler is shown in Fig.4.12. In this case the robot can track the mobile target very well. The average tracking error throughout the whole experiment is only 0.51 mm. It is still worth noting that this method is not applicable to the system described in Section 4.2. The relevant results

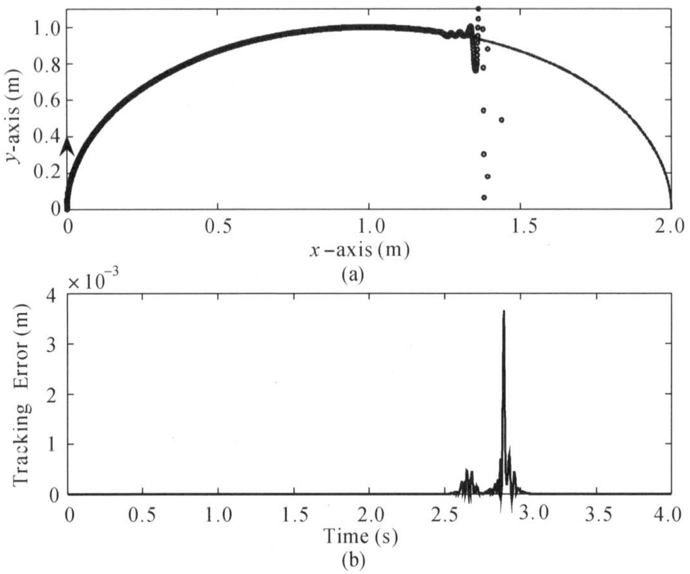

Fig. 4. 10 Target tracking performance under open-loop scheduling
(a) Mobile Target Tracking; (b) Tracking Error

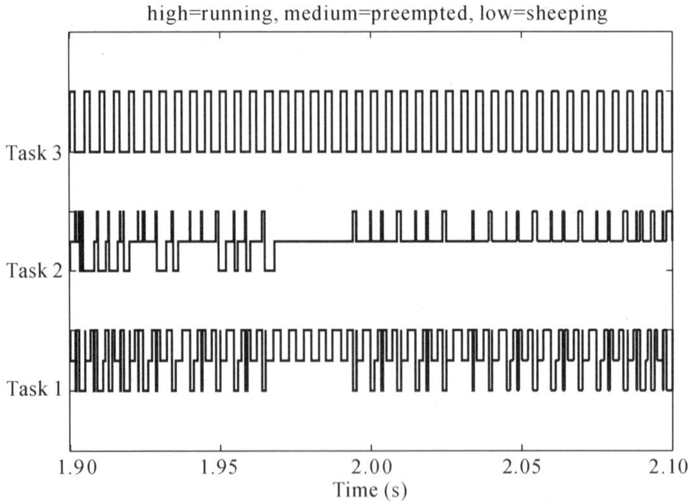

Fig. 4. 11 CPU schedule under open-loop scheduling

are given here to compare the fuzzy feedback scheduling method against the ideal one.

The execution of all tasks around time $t = 2$ s is depicted in Fig.4.13, where FS denotes the feedback scheduler task. It is not hard to find out that the ideal feedback scheduler dynamically adapts periods of control tasks τ_1 and τ_2, which correspond to the control loops' sampling periods. The changes in task periods can be observed

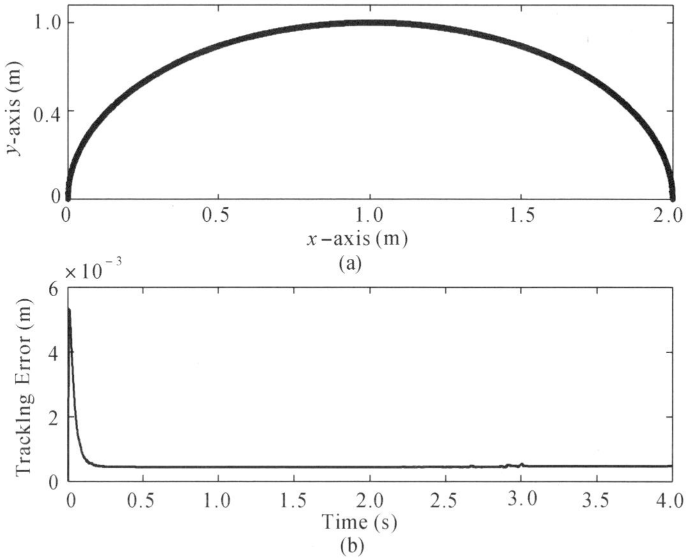

Fig. 4. 12 Target tracking performance under ideal feedback scheduling (a) Mobile Target Tracking; (b) Tracking Error

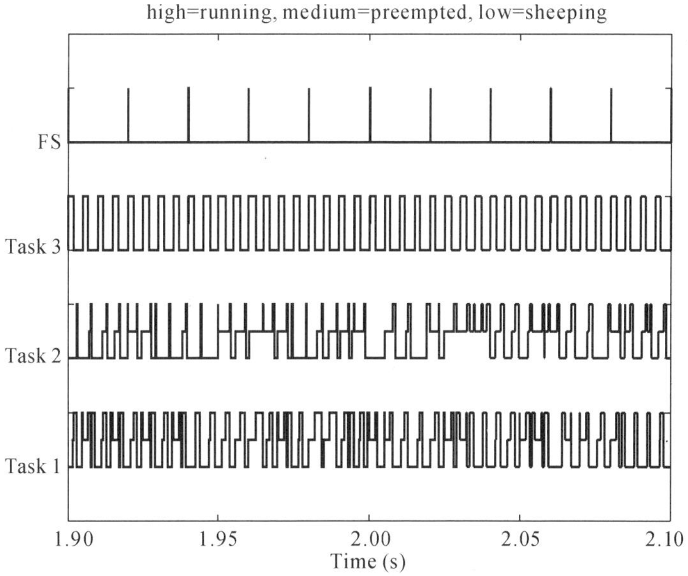

Fig. 4. 13 CPU schedule under ideal feedback scheduling

from the time interval between two consecutive executions. The situation in which task τ_2 is continuously preempted is avoided, which in contrast happens in the case of open-loop scheduling. Sufficient CPU time has been obtained by each task.

By comparing Fig.4.12 with Fig.4.10, it can be seen that feedback scheduling is efficient in dealing with dynamic variations of system workload. It thus provides control performance guarantees for systems operating under changing conditions.

To assess the performance of fuzzy feedback scheduling, extensive simulations have been done for systems with measurement noises of different magnitudes. Some representative results are depicted in Fig.4.14. The moving tracks of the robot in these cases are the same as in the case of ideal feedback scheduling, and hence are omitted here. It can be seen that fuzzy feedback scheduling is capable of handling the negative impact of measurement noises. It is robust and adaptable to noises of different magnitudes. For instance, the robot tracks the target considerably well even when the standard deviation of the white Gaussian noise reaches 0.3, with a small average tracking error of 0.53 mm.

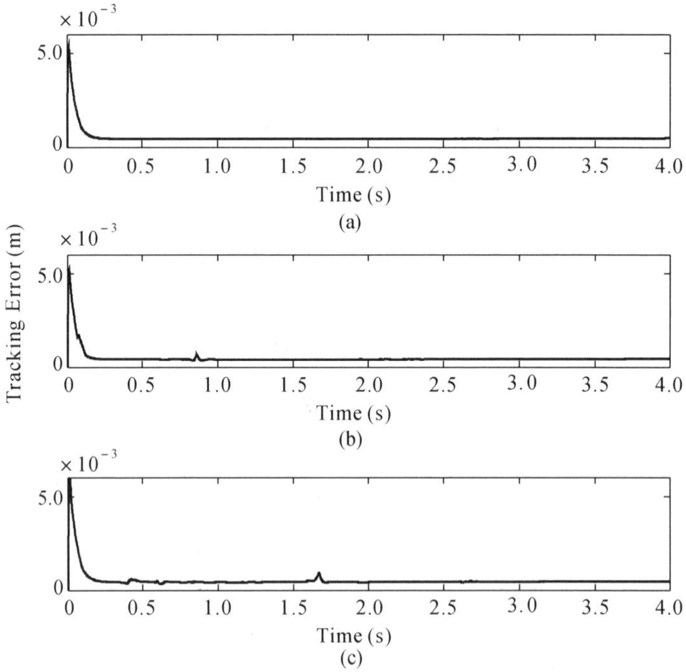

Fig. 4.14 Target tracking performance under fuzzy feedback scheduling (a) $r = 0.1$; (b) $r = 0.2$; (c) $r = 0.3$

Fig.4.15 gives the CPU schedule for $r = 0.1$. It can be seen that the periods of two control tasks are changed at runtime. Just as the ideal feedback scheduler does, the fuzzy feedback scheduler successfully prevents the lowest-priority task from being continuously preempted. The available CPU time is allocated in a reasonable way. This mainly benefits from the dynamic adaptation of task periods, see Fig.4.16 that depicts the task periods for the case $r = 0.1$.

The CPU workload (i. e. total requested CPU utilization) under different schemes is shown in Fig.4.17. In the case of open-loop scheduling, the system workload varies with each task execution time due to fixed task periods. In this

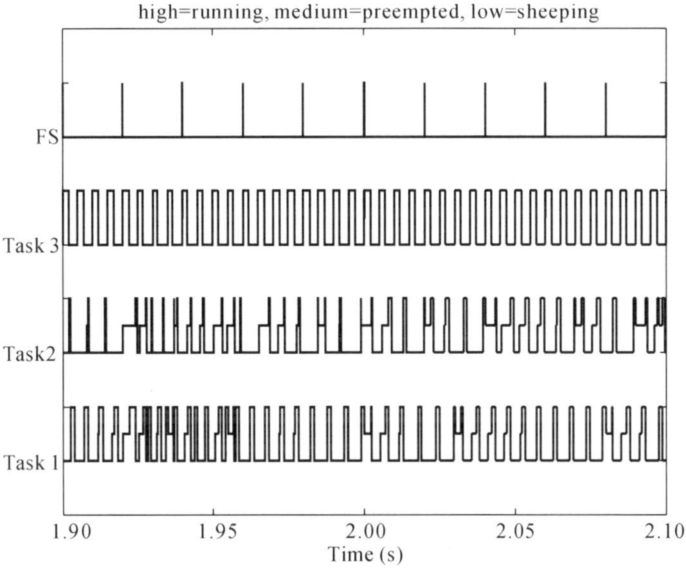

Fig. 4. 15 CPU schedule under fuzzy feedback scheduling

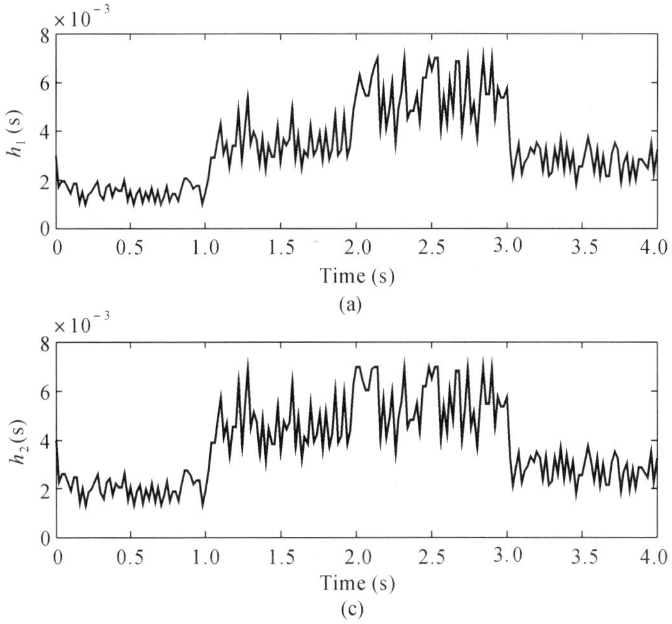

Fig. 4. 16 Task periods under fuzzy feedback scheduling
(a) h_1; (b) h_2

context the factors that affect the level of workload include the changes in tasks' average execution time and the random process ε within task execution time. The system is overloaded throughout the time interval 2 – 3 s. This incurs quite a large number of deadline misses. Therefore, the control loop for the y coordinate goes unstable, as mentioned above. When the ideal feedback scheduler is used, the (requested) CPU utilization remains at the desired level 85% almost all the time. As a consequence, the system schedulability is respected. The deviation of the actual CPU utilization from the desired level that appears several times in the simulation is due to the constraint on the maximum allowable sampling periods for control loops. In the context of fuzzy feedback scheduling (where $r = 0.1$), the requested CPU utilization keeps around the desired level. In a sense, Fig.4.17 explains the main reason for the different performances of the system under different schemes.

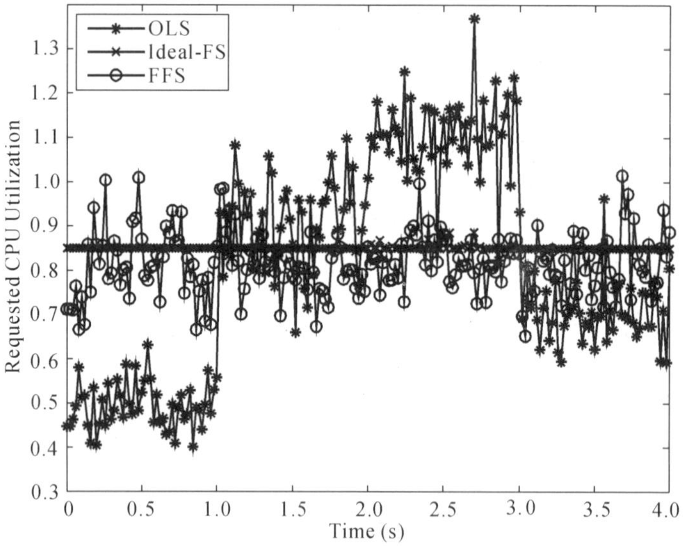

Fig. 4. 17 CPU workload under different schemes

Fig.4.18 gives the accumulated tracking error $\int_0^t ERR_{\text{track}}(\tau)\,d\tau$. In cases where $r \leqslant 0.2$, the curves of the accumulated tracking error under fuzzy feedback scheduling are almost identical with that under ideal feedback scheduling, and therefore they are not shown in Fig.4.18. Fuzzy feedback scheduling and ideal feedback scheduling deliver almost the same performance for $r = 0.3$.

It can be outlined from the above results that: 1) fuzzy feedback scheduling is capable of coping with unavailability of task execution time and uncertainty in system workload; 2) it is robust to measurement noises of different magnitudes; 3) it can achieve system performance comparable to that of ideal feedback scheduling.

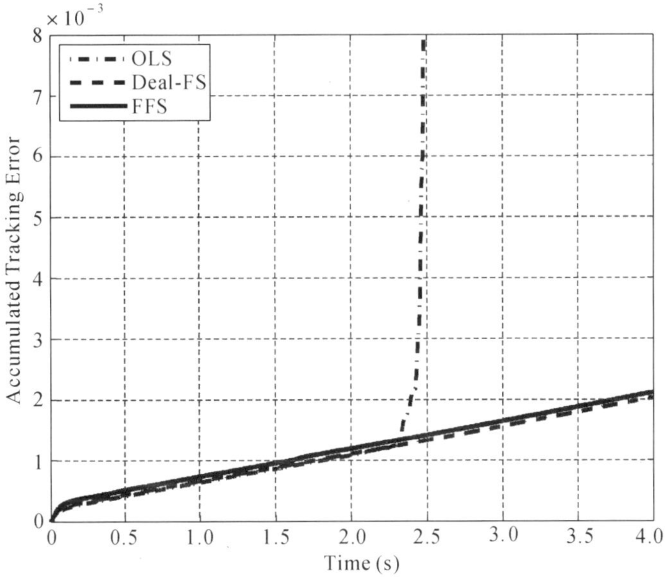

Fig. 4.18 Accumulated tracking error

4.6 Summary

In some real-world applications, it is hard, if not impossible, to gather online the tasks' temporal attributes such as execution time. The presence of measurement noises also causes available system parameters to be imprecise. In the absence of some necessary information, conventional feedback scheduling schemes may become unfeasible. Furthermore, most often they do not pay much attention to measurement noises, which may negatively impact the feedback scheduling performance.

This chapter deals with the unavailability of task execution time and the presence of measurement noises in embedded control systems. By incorporating the fuzzy control technology into the framework of feedback scheduling, a fuzzy feedback scheduling scheme is presented. Two alternative architectures for implementing the scheme are discussed. A simplified mobile robot system is used for a concrete case study. The design procedures of the fuzzy feedback scheduler are detailed. Using the look-up table method, an intelligent feedback scheduling algorithm is obtained that delivers low computational overhead and is easy to implement. What makes the fuzzy feedback scheduling scheme particularly interesting is that it does not rely on the availability of task execution time like conventional feedback scheduling methods. Simulation results highlight the good performance of fuzzy feedback scheduling.

References

[AMI05] M. Amirijoo, J. Hansson, S. Gunnarsson, S. H. Son. Enhancing Feedback Control Scheduling Performance by Online Quantification and Suppression of Measurement Disturbance. *Proc. 11th IEEE RTAS*, California, USA, pp.2 – 11, 2005.

[AND05] M. Andersson, D. Henriksson, A. Cervin, K.E. Årzén. Simulation of Wireless Networked Control Systems. *Proc. of 44th IEEE Conf. on Decision and Control and European Control Conf.*, Seville, Spain, pp.476 – 481, 2005.

[BUT02] G. Buttazzo, L. Abeni. Adaptive Workload Management through Elastic Scheduling. *Real-Time Systems*, Vol. 23, No.1/2, pp.7 – 24, 2002.

[CER02] A. Cervin, J. Eker, B. Bernhardsson, K.E. Årzén. Feedback-Feedforward Scheduling of Control Tasks. *Real-Time Systems*, Vol. 23, No.1, pp.25 – 53, 2002.

[DIA02] Y. Diao, J.L. Hellerstein, S. Parekh. Optimizing Quality of Service Using Fuzzy Control. *Proc. 13th IFIP/IEEE DSOM*, Springer-Verlag, pp.42 – 53, 2002.

[JIN04] H. Jin, H.A. Wang, Y. Fu, Q. Wang, H. Wang. A Fuzzy Feedback Control Real-Time Scheduling Algorithm. *Chinese Journal of Software*, Vol.15, No.6, pp.791 – 798, 2004.

[KEV98] K.M. Passino, S. Yurkovich. *Fuzzy Control.* Addison-Wesley, 1998.

[LEI05] F. Lei, Y. Zou, J. Zhou. A Fuzzy Feedback Scheduling System for Real-Time Control Tasks. *Chinese Journal of Information and Control*, Vol. 34, No.5, pp.585 – 590, 2005.

[MAR02] P. Martí. Analysis and Design of Real-Time Control Systems with Varying Control Timing Constraints. Ph.D Thesis, Technical University of Catalonia, 2002.

[REN02] F. Ren, Y. Ren, X. Shan. Design of a Fuzzy Controller for Active Queue Management. *Computer Communications*, Vol. 25, No.9, pp. 874 – 883, 2002.

[SIM05] D. Simon, D. Robert, O. Sename. Robust Control/Scheduling Co-Design: Application to Robot Control. *Proc. IEEE RTAS*, California, USA, 2005.

[SUN97] Z. Sun, Z. Zhang, Z. Deng. *Theory and Technology of Intelligent Control*, Tsinghua University Press, Beijing, 1997.

[SUN06] Y.X. Sun, J. Chu. *Industrial Process Control Technology*, Beijing: Chemical Industry Press, 2006.

[WEN01] X. Wen, L. Zhou, D. Li, C. Bei. *Analysis and Applications of Matlab Fuzzy Logic Toolbox*, Beijing: Science Press, 2001.

[XIA05a] F. Xia, L.P. Liu, Y.X. Sun. Flexible Quality-of-Control Management in Embedded Systems Using Fuzzy Feedback Scheduling. *Lecture Notes in Artificial Intelligence*, Vol. 3642, pp.624 – 633, 2005.

[XIA05b] F. Xia, X.F. Shen, L.P. Liu, Z. Wang, Y.X. Sun. Fuzzy Logic Based Feedback Scheduler for Embedded Control Systems. *Lecture Notes in Computer Science*, Vol. 3645, 2005, pp.453 – 462.

[XIA07] F. Xia, Y.X. Sun, Y.C. Tian, M.O. Tadé, J.X. Dong. Fuzzy Feedback Scheduling of Resource-Constrained Embedded Control Systems, Submitted, 2007.

[ZOU03] Y. Zou. Research on Scheduling for Open Real-Time Systems. Ph.D Thesis, Institute of Software, The Chinese Academy of Sciences, 2003.

PART III

ENERGY MANAGEMENT

Chapter 5

Energy-Aware Feedback Scheduling

Energy consumption plays an ever-increasingly important role in various embedded systems, especially when the system is battery-powered. This chapter studies the problem of dynamic energy management in embedded control systems that operate in dynamic environments. The aim is to reduce the energy consumption of the processor as much as possible, while preserving the required QoC.

To cope with practical problems such as CPU workload variations and unpredictability of task execution time, a feedback control real-time scheduling methodology called Energy-Aware Feedback Scheduling (EAFS) is suggested. This approach combines the management of both energy consumption and QoC. By exploiting the Dynamic Voltage Scaling (DVS) technique, it indirectly adapts tasks' execution time through dynamically adjusting CPU speed. In this way, the CPU utilization is maintained at a desired high level, and closed loop power management is achieved. After analytically modelling the DVS system, a control-theoretic design and analysis method for the feedback scheduler is proposed, which delivers predictable feedback scheduling performance. The performance of EAFS is assessed by extensive simulation experiments. Its merits are highlighted by means of comparisons with other related schemes.

5.1 Introduction

Recent years have witnessed the rapid evolution and widespread applications of mobile wireless embedded devices/platforms in computer and communication engineering systems. As the miniaturization of physical size continues, an increasing number of embedded devices now utilize batteries as power sources [PIL01, UNS03]. Some examples are PDAs, laptop and pocket computers, etc. In addition, real-time control applications built upon battery-powered computing platforms are of paramount importance for many areas such as robotics, avionics, and deep-space discovery.

It could be envisioned that with the burst of pervasive computing and communication techniques, the growth of battery-powered embedded devices applied to

Real-Time Control Systems will continue into the future. Battery lifetime thus becomes a critical factor that determines the usability of the system, and therefore must be taken into account during system design. One straightforward approach to prolong battery lifetime is to improve battery capacity. However, the evolution of battery technology cannot keep pace with the ever-increasing energy demands of the applications, and no dramatic breakthrough is foreseeable right now [ZHU05].

Consequently, control engineers are confronted with a "new" kind of resource management problem, i.e. energy management. Since the energy budget is limited with a given battery technology, the energy consumption should be reduced as much as possible to extend battery life. On the other hand, the processing capability of modern computing platforms has become much more powerful in recent years, in response to the increase of application requirements. More energy is consequently demanded. In general, high-quality system performance and low energy consumption are at odds with each other [PIL01]. Therefore, in energy-limited embedded controllers it is necessary to manage simultaneously both QoC and energy consumption.

Even when the available energy is not severely constrained, saving energy in Real-Time Control Systems still has great benefits. For instance, dataquest has reported that worldwide total power dissipation of processors in PCs was 160 MW in 1992, and went up to 9000 MW by 2001 [UNS03]. This value is still growing. The practical significance of saving energy cannot be overly emphasized. In fact, low power has become one of the most important features of embedded processors and has significant impact on the operational cost as well as the environment in which we live [ZHU05]. Low power design has therefore become a technical trend in almost all applications where energy consumption is involved.

In this part of energy management (Chapters 5 and 6), our attention is focused on energy consumption of embedded processors in Real-Time Control Systems. In most embedded systems, the processor is the most energy-consuming component. This is true on large laptop computers for example those have many components including large displays with backlighting, as well as on small handheld devices that have fewer components [KER05, PIL01]. In applications where the processor may not be the most energy-consuming component, it is also essential to efficiently manage processor power in order to reduce the overall system energy consumption.

The vast majority of modern processors are built using the prevailing CMOS technology. There are basically three components of power dissipation in CMOS circuits [UNS03]: dynamic, static and short-circuit. Dynamic power is dissipated due to the switching of gates. Static power is mainly due to leakage current between the power supply and the ground. Short-circuit power is dissipated when both the negative metal-oxide-semiconductor and positive metal-oxide-semiconductor transistors are conducting simultaneously, and is comparatively insignificant. Among these three components, dynamic power contributes to the largest percentage of the total power consumption in current processor design. Though static power consumption is expected to increase in the future, it can be ignored under most circumstances. Therefore only dynamic power consumption of the CPU is considered in this book.

5.1.1 Motivation

Most traditional task scheduling policies are originally designed for real-time systems that have fixed CPU speed, and hence cannot be directly applied to managing the processor power of embedded controllers. Dynamic voltage scaling (also known as dynamic voltage and frequency scaling) [AYD04, KER05, PIL01, WEI94] is currently the most promising energy saving technique. DVS exploits the convex (normally quadratic) relationship between CPU energy consumption and voltage. Its basic idea is to minimize CPU idle time in order to cut the CPU energy consumption, while satisfying timing constraints of real-time tasks. Most existing microprocessors such as Intel's Xscale and StrongARM, AMD's K6 − 2 +, and Transmeta's Crusoe support this technique [ZHA06a, ZHA06b]. DVS takes advantage of the hardware characteristics of these processors to reduce energy consumption through simultaneously changing the supply voltage and operating frequency with respect to system workload variations. It has been demonstrated to be highly effective in saving energy for different types of applications, both real-time and non real-time.

For CMOS-based processors, the dynamic energy consumption can be approximated in most cases by [KWO05]:

$$E_{\mathrm{total}} = \sum_{i=1}^{N} E_i = \sum_{i=1}^{N} (R_i \cdot C_i \cdot V_{\mathrm{dd}}^2) \propto V_{\mathrm{dd}}^2, \tag{5.1}$$

where N is the number of concurrent tasks, R_i is the total number of cycles required for execution of the task, C_i is a constant indicating the average switched capacitance per clock cycle, and V_{dd} is the supply voltage. For a given workload, the CPU energy consumption is proportional to the square of the supply voltage. Lowering supply voltage will yield a quadratic reduction in CPU energy consumption, which is the fundamental principle that makes DVS a possibility.

However, lowering the supply voltage also increases the circuit delay T_{d}. Their relationship approximates that [AYD04, MEJ04, PIL01]:

$$T_{\mathrm{d}} \propto \frac{V_{\mathrm{dd}}}{(V_{\mathrm{dd}} - V_{\mathrm{t}})^2}, \tag{5.2}$$

where V_{t} is the threshold voltage. It can be seen that the circuit delay increases with decreasing supply voltage. Furthermore, the circuit delay is in approximately inverse proportion to the supply voltage. For control systems, a natural result of the increase in circuit delay is longer control delay, which in turn degrades control performance. This problem will be more pronounced if the task schedulability is violated due to prolonged task execution time. It is clear that there is a fundamental trade-off between saving energy and improving control performance when applying DVS to embedded controllers.

While significant researches and development efforts have been made on DVS in many areas such as general-purpose computing systems, multimedia, wireless sensor networks, etc., very little work has been done on energy management in the context of real-time control. The application of feedback scheduling to this area is also unexplored. The reader is referred to Chapter 1 for details on related work.

5.1.2 Contributions

This chapter deals with energy constraint in multitasking embedded Real-Time Control Systems. The objective is to reduce the CPU energy consumption as much as possible, while preserving QoC guarantees. For this purpose, a feedback control real-time scheduling scheme [XIA05a, XIA07] that integrates energy management and QoC management is presented. It is called Energy-Aware Feedback Scheduling (EAFS). This approach is based on the DVS technology, and realizes closed loop energy management in the framework of feedback scheduling.

In practice, the execution time of a control algorithm is usually data dependent. That is, the execution time of a control task may vary with different sampled data that indicate different states of the controlled system. For control systems that use hybrid control techniques (e. g. [CER00]), multi-version controllers supporting multi-level services (e. g. [LIN04]), or anytime control algorithms that can be modelled using the imprecise computation framework (e.g. [XIA05a, XIA05c]), it is considerably hard, if not impossible, to obtain (or even estimate) the accurate execution time of control tasks. This problem is further accentuated in systems where COTS components and non real-time operating systems are deployed. In these situations, the effectiveness of traditional open scheduling algorithms and WCET-based DVS schemes becomes doubtful.

To reduce CPU energy consumption more efficiently, we propose to design the feedback scheduler intended for DVS-enabled systems by exploiting discrete-time control theory. With this feedback scheduler, the supply voltage and operating frequency are dynamically adjusted, and all tasks' execution time will accordingly be changed. The CPU utilization will be controlled at a desired level. In most cases quite a high level of the setpoint for CPU utilization can be specified. This delivers at least two advantages:

(1) Since the CPU utilization level remains considerably high most of the time, the CPU idle time is largely reduced, which benefits energy consumption reduction;

(2) Task schedulability is preserved when the CPU utilization keeps at the desired level, and therefore the QoC guarantee is provided.

Simulation experiments are conducted to demonstrate the capability of the proposed approach with respect to dealing with practical problems such as CPU workload variations and unpredicatable task execution time. The results show that the proposed approach can dramatically reduce CPU energy consumption, while guaranteeing the QoC of the system.

In summary, the major contributions of this chapter are as follows:

(1) A novel feedback scheduling scheme is suggested for DVS-enabled embedded controllers that operate in unpredictable environments. It can reduce CPU energy consumption and guarantee QoC at the same time.

(2) An analytical model for the DVS system is developed. Based on this model, a control-theoretic method is proposed for feedback scheduler design and analysis. This method enhances the predictability of the resulting performance of the feedback scheduler.

(3) Extensive simulation experiments are carried out, with favourable results in comparison the proposed approach with several typical solutions.

The organization of this chapter is as follows. In Section 5.2 the system model and the energy consumption model used in this chapter are described. Section 5.3 details the Energy-Aware Feedback Scheduling scheme. Following the basic idea of this scheme, the DVS system considered is modelled analytically. A feedback scheduler is then designed using feedback control theory. A preliminary stability analysis method for the feedback scheduling system is also given. In Section 5.4, simulation experiments are conducted to assess the performance of the proposed approach in comparison with other schemes. Section 5.5 concludes this chapter.

5.2 Problem Statement

5.2.1 System Model

Consider an energy-limited variable voltage processor on which N independent control tasks run concurrently. Each control task is responsible for controlling an independent physical process. The system supports the DVS technology. Assume for simplicity that the voltage/frequency of the CPU can be adjusted continuously with a scaling factor $\alpha \in [\alpha_{\min}, 1]$. Since the clock frequency of CPU is (approximately) proportional to the supply voltage, α will also be used to denote the CPU speed. Notice that α is a normalized value equal to the ratio of actual CPU operating speed to the maximum possible speed. For example, for an Intel® Xscale[1] processor with a maximum operating voltage of 1.8 V, it holds that $\alpha = 1.0/1.8 = 0.556$ when the actual supply voltage is set to 1.0 V.

The timing parameters of each control task i are described as follows:

(1) h_i: period, which is equal to the sampling period of the control loop i, and is fixed during runtime.

(2) $\hat{c}_{i,\text{nom}}$: estimated execution time at full CPU speed (i. e. the maximum possible speed). For simplicity it is called *estimated nominal execution time*.

(3) $\hat{c}_i$: estimated execution time at actual CPU speed associated with α, which satisfies $\hat{c}_i = \hat{c}_{i,\text{nom}}/\alpha$.

(4) $c_{i,\text{nom}}$: actual execution time at full CPU speed, which is variable and unpredictable at runtime. Assume that $c_{i,\text{nom}} = \lambda \hat{c}_{i,\text{nom}}$, where λ is a variable *execution time factor*.

---

1 http://www.intel.com/design/intelxscale/

(5) c_i: actual execution time at actual CPU speed, which satisfies $c_i = c_{i,\text{nom}}/\alpha = \lambda\, \hat{c}_i$. It is clear that c_i changes with $c_{i,\text{nom}}$, and is also unpredictable.

For the sake of simple description, the following definitions are used in this chapter:

(1) CPU utilization $U = \sum \frac{c_i}{h_i}$. Accordingly, the estimated CPU utilization $\hat{U} = \sum \frac{\hat{c}_i}{h_i}$.

(2) CPU workload $\omega = U \cdot \alpha = \sum \frac{c_{i,\text{nom}}}{h_i}$, and the estimated CPU workload $\hat{\omega} = \hat{U} \cdot \alpha = \sum \frac{\hat{c}_{i,\text{nom}}}{h_i}$.

Although different kinds of real-time task scheduling policies can be employed, we restrict ourselves in this chapter to illustrate the proposed approach based on the EDF algorithm. According to the well-known schedulability utilization bound for EDF (see Theorem 2.3), the schedulability condition associated with the processing speed α can be expressed by:

$$\sum_{i=1}^{N} \frac{c_i}{h_i} \leqslant 1 \quad \Leftrightarrow \quad \omega \leqslant \alpha . \tag{5.3}$$

Because $\alpha_{\min} \leqslant \alpha \leqslant 1$, it is assumed that $\omega = \sum \frac{c_{i,\text{nom}}}{h_i} \leqslant 1$ such that feasible solutions exist under all circumstances. In this monograph, the switching overheads, including both energy overhead and time overhead, between different voltage and frequency levels are neglected. This is based on the observation that the switching time of prevailing processors is always negligibly small in comparison with task periods. For instance, the time overhead of a Transmeta Crusoe[2] processor for switching voltage per step is less than 20 μs [MEJ04], while the periods of control tasks are generally at least on the order of ms.

5.2.2 Energy Model

In the system considered, the energy expenditure of the processor is sampled with a fixed time interval. In this way, the whole CPU energy consumption can be related to the energy expenditure per sample as a function of the normalized processing speed α. [In GUT97], the energy consumption per sample for first-order CMOS delay models has been derived, which is given by:

$$E(\alpha) = C_0 V_0^2 T_s f_{\max} \alpha \left[\frac{V_t}{V_0} + \frac{\alpha}{2} + \sqrt{\alpha \frac{V_t}{V_0} + \left(\frac{\alpha}{2}\right)^2} \right]^2 , \tag{5.4}$$

where C_0 is the average switched capacitance, V_t is the device threshold voltage, V_0

---

2 http://www.transmeta.com/

$= (V_{max} - V_t)^2/V_{max}$, T_s is the sampling interval, and f_{max} is the maximum clock speed. For a given DVS system, it has been shown [SIN01] that Eq.(5.4) can be equivalently approximated by a simple quadratic model:

$$E(\alpha) = \alpha^2. \tag{5.5}$$

In this chapter Eq.(5.5) is used to calculate the normalized energy consumption of the CPU. With this model, it is easy to understand that α should be minimized in order to maximize energy saving. Ideally, the minimum possible CPU speed under task schedulability constraint can be obtained according to Eq.(5.3), which turns to be $\max\{\omega, \alpha_{min}\}$. In this case, the energy consumption will be minimized if the CPU speed is set to this level. Because ω is unknown at runtime, however, the exact minimum CPU speed is impossible to deduce as in the ideal case. Therefore, methods are needed to handle the uncertainties in task' execution time.

5.3 Energy-Aware Feedback Control Scheduling

Feedback control theory is undoubtedly one of the most powerful tools for dealing with uncertainties in various engineering systems. It has been successfully applied to real-time CPU scheduling, e.g. [LU02, SOR05]. In this section a novel feedback scheduling scheme with energy consideration is presented. The feedback scheduler will be designed using feedback control techniques. This leads to predictable performance of the feedback scheduling system.

5.3.1 Basic Idea

Using the general framework of feedback scheduling depicted in Fig. 2.10, we propose to treat the DVS-enabled multitasking embedded controller as the controlled process within the feedback scheduling loop. The feedback scheduler serves as a controller in view of feedback control. The choice of key control-related variables is discussed below.

The *controlled variable* is chosen to be the actual CPU utilization. It is well-known with control engineers that deadline miss deteriorates the QoC. Consequently, deadline miss should be avoided in favour of QoC improvement. It is then unsuitable to use the deadline miss ratio as the controlled variable. On the other hand, as long as the CPU utilization is maintained at a considerably high level without exceeding the upper bound of the schedulability condition, the QoC will then be guaranteed. Meanwhile, the idle time of the CPU will also be minimized, which implies reduced energy consumption.

The *manipulated variable* is set to be the CPU speed α. This is quite intuitive and is easy to understand, since the CPU speed seems to be the only factor that directly determines power dissipation and, at the same time, effects on control performance. The operating speed of CPU will be adjusted each time the feedback scheduler runs, and remain fixed till the next invocation of the feedback scheduler.

That is, a system-level dynamic DVS technique is exploited. Similar to general control applications, the purpose of manipulating α is to drive the controlled variable i.e. the CPU utilization to settle down at a desired level. From a perspective of timing parameters, the adjustment of the CPU speed can be equivalently regarded as the regulation of execution time of control tasks. Moreover, it is the task execution time rather than the CPU speed that affect the QoC directly. From this insight, the tasks' execution time can be viewed as the manipulated variable.

The *setpoint* U_R is the desired CPU utilization level. In order for CPU time to be fully used and the energy expenditure to be reduced as much as possible, a higher (desired) level of CPU utilization will always be better. At the best, the actual CPU utilization will keep at the upper bound of task schedulability conditions, which is 100% for EDF. A result of this ideal case is that the lowest possible power dissipation is achieved while the system schedulability is also preserved. In real applications, due to the uncertainty in task execution time, the possibility of missing deadlines will increase if U_R approaches 100% too closely. As a consequence, the control performance may be degraded. On the other hand, if U_R is low, some resources will be wasted, which affects the effectiveness of energy saving. Therefore, in practice an appropriate U_R value should be chosen on the basis of knowledge about, e.g. the magnitudes of actual variations in the tasks' execution time.

Since the feedback scheduler is time-triggered with a fixed invocation interval of T_{FS}, the DVS technique employed here is naturally interval based [WEI94]. The system periodically adjusts the operating speed of the CPU within the embedded controller. During every invocation interval all tasks run upon the same CPU speed. The feedback scheduling method developed in this chapter is unique in that it depends on the adjustment of execution time. That is, the scheduling parameters that the feedback scheduler adapts are the tasks' execution time. In contrast, Chapters 3 and 4 exploit methods that rely on period adjustment.

Remark 5.1 Because of the inherent uncertainty and unpredictability of the tasks' execution time, the proposed approach does not provide hard real-time guarantees. In some special cases, for example, when the workload increases abruptly, the phenomenon of deadline miss may appear. However, these deadline misses will not significantly jeopardize the QoC in general cases. This is because: 1) the control-theoretic design methodology can enforce the predictability of the resulting feedback scheduling performance, and hence special scenarios such as abrupt workload increase can be effectively handled, thus avoiding severe deadline misses; 2) practical control applications are usually designed with the capability of tolerating some deadline misses. Of course, deadline miss could potentially cause instability if it is too severe to be tolerable for a control system. This must be taken into account during the course of feedback scheduler design.

5.3.2 Modelling

As a prerequisite of using feedback control techniques, a mathematical model must be established for the DVS system. For this purpose, examine the following calculation of the CPU utilization in the time interval $[jT_{FS}, (j+1)T_{FS}]$:

$$U(j+1) = \sum_{i=1}^{N} \frac{c_i(j)}{h_i} = \lambda(j) \sum_{i=1}^{N} \frac{\hat{c}_i(j)}{h_i} = \frac{\lambda(j)}{\alpha(j)} \sum_{i=1}^{N} \frac{\hat{c}_{i,\,\mathrm{nom}}(j)}{h_i}. \tag{5.6}$$

In the above equation, $U(j+1)$ is the output of the feedback scheduling system, $\alpha(j)$ is the control input from a control perspective, $\lambda(j)$ is the execution time factor, and the number of control loops N and the estimated nominal execution time $\hat{c}_{i,\,\mathrm{nom}}$ are known. It should be noted that N and $\hat{c}_{i,\,\mathrm{nom}}$ may change over time.

Since the estimated execution time of different jobs of a task may be different even in the same feedback scheduler invocation interval, the mean of the estimated execution time of all jobs associated with each task can be used as $\hat{c}_{i,\,\mathrm{nom}}$ in every interval. For simple description, assume that the estimated execution time of all jobs of a task in the same interval is equal to $\hat{c}_{i,\,\mathrm{nom}}(j)$.

Since the variability of $\lambda(j)$ could complicate the design of the feedback scheduler, a simplification method is used in modelling. To guarantee stability in all circumstances, the execution time factor $\lambda(j)$ in Eq.(5.6) is replaced by its maximum possible value $K_\lambda = \max\{\lambda(j)\}$. The same method has been used in [LU02] for modelling the CPU scheduling system. With $\hat{\omega}(j) = \sum_{i=1}^{N} \frac{\hat{c}_{i,\,\mathrm{nom}}(j)}{h_i}$, Eq.(5.6) can be rewritten as:

$$U(j+1) = \frac{K_\lambda\, \hat{\omega}(j)}{\alpha(j)}. \tag{5.7}$$

Obviously, the system output $U(j+1)$ is in inverse proportion to $\alpha(j)$. To achieve a linear model, let $\beta(j) = 1/\alpha(j)$. Then the following formula is obtained:

$$U(j+1) = K_\lambda\, \hat{\omega}(j)\beta(j). \tag{5.8}$$

In Eq.(5.8), $\hat{\omega}(j)$ may vary during runtime, though it is known. Strictly speaking, this system is a time variant system with variable parameters. One possible approach to dealing with the variations of $\hat{\omega}(j)$ is just the same as what we have done with $\lambda(j)$, that is, to use the maximum possible value of $\hat{\omega}(j)$ to replace it. However, unlike $\lambda(j)$ that is unknown, $\hat{\omega}(j)$ is known to the feedback scheduler. In this context, an online gain scheduling method [SIM05] is used to compensate for the dynamic variations of $\hat{\omega}(j)$, see Fig.5.1. Accordingly the term $\hat{\omega}(j)$ is dropped from the DVS system model Eq.(5.8). After performing z-transform on Eq.(5.8), the following discrete-time model is obtained for the DVS system:

$$G_P(z) = \frac{U(z)}{\Delta\beta(z)} = \frac{K_\lambda}{z-1}, \tag{5.9}$$

where $\Delta\beta(j) = \beta(j) - \beta(j-1)$.

Remark 5.2 Both the CPU utilization U and the variable β are subject to saturation. The ranges of their possible values are: $0 \leqslant U \leqslant 1$, $1 \leqslant \beta \leqslant 1/\alpha_{\min}$.

To this end, an analytical model for the DVS system has been deduced. Described below is how to design the feedback scheduler based on this model and taking advantage of feedback control techniques.

5.3.3 Feedback Scheduler Design

From the viewpoint of feedback control, the model given in Eq.(5.9) is quite simple. In theory, a lot of well-established control techniques can be employed to design the controller, i.e. the feedback scheduler in the framework of feedback scheduling. As a simple yet representative illustration, the PI control algorithm is adopted here. The architecture of the feedback scheduling loop is shown in Fig.5.1.

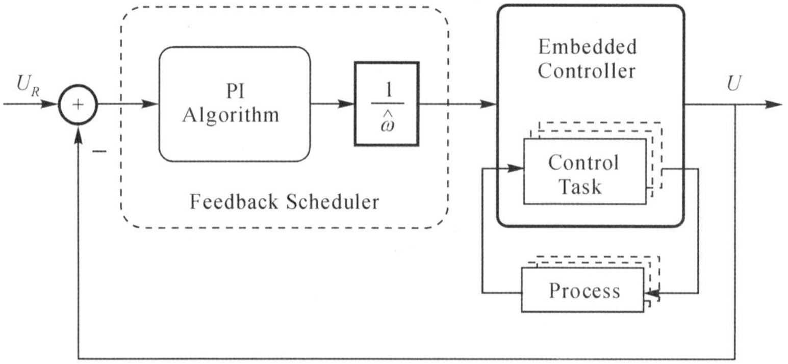

Fig. 5.1 Schematic structure of Energy-Aware Feedback Scheduling

While many other control algorithms are also suitable for constructing the feedback scheduler, it is still worth explaining the reasons for a nontrivial question: why a PI algorithm? Some reasons are given below.

(1) The model given in Eq.(5.9) represents a first-order system. It is not hard to design an effective controller for such a system. Therefore, provided that the performance requirements are met, the feedback scheduler should be simplified as much as possible to minimize the feedback scheduling overhead.

(2) PID is the most widespread control algorithm in practical control applications. It is well suited for low order dynamical systems, and is easy to implement.

(3) The derivative component of the PID algorithm may amplify the effect of noise, and in consequence is excluded. An additional benefit of not using PID but PI is that this reduces not only the complexity of offline design bus also the online computational overhead of the feedback scheduler.

As mentioned above, the dynamic gain scheduling component $1/\hat{\omega}(j)$ is for compensating for the variable yet known term $\hat{\omega}(j)$ in Eq.(5.8). For the purpose of feedback scheduler design, the system model Eq.(5.9) is utilized, and the gain scheduling component is ignored.

Regarding the determination of the coefficients associated with the PI control algorithm, the pole placement method is employed, since it is widely used in the control community. In this way, predictable performance of feedback scheduling can be achieved explicitly.

The discrete-time transfer function of a PI controller is given by

$$G_C(z) = \frac{\Delta\beta(z)}{\Delta U(z)} = K_P + K_I \frac{z}{z-1}.$$

Combining it with Eq.(5.9) in the framework of Fig.5.1 gives the closed loop transfer function of the feedback scheduling loop:

$$G(z) = \frac{G_C(z)G_P(z)}{1 + G_C(z)G_P(z)} = \frac{\left(K_P + K_I \frac{z}{z-1}\right)\frac{K_\lambda}{z-1}}{1 + \left(K_P + K_I \frac{z}{z-1}\right)\frac{K_\lambda}{z-1}} \qquad (5.10)$$

$$= \frac{K_\lambda(K_P + K_I)z - K_\lambda K_P}{z^2 + (K_\lambda K_P + K_\lambda K_I - K_\lambda - 2)z + 1 - K_\lambda K_P}$$

Let $a \pm b_i$ be the desired closed loop poles. The corresponding characteristic equation is:

$$(z - a - b_i)(z - a + b_i) = 0, \qquad (5.11)$$

i.e.

$$z^2 + 2az + a^2 + b^2 = 0. \qquad (5.12)$$

According to the principle of pole placement, the following equation group is obtained from Eqs.(5.10) and (5.12):

$$\begin{cases} K_\lambda K_P + K_\lambda K_I - 2 = 2a \\ 1 - K_\lambda K_P = a^2 + b^2 \end{cases}. \qquad (5.13)$$

Once the desired closed loop poles are chosen, the control coefficients K_P and K_I can be obtained by simply solving Eq.(5.13). Thus the feedback scheduler composed of the PI algorithm can be designed accordingly.

Now that the feedback scheduler has been designed using feedback control techniques, control theoretic methods can also be exploited to analyze the resulting performance, stability for example, of the feedback scheduling loop. Using well-established results in the field of discrete-time control [AST97], it is not difficult to understand the following necessary and sufficient condition for feedback scheduling system stability.

Theorem 5. 1 A feedback scheduling system designed using the above approach is stable if and only if the closed loop poles $a \pm b_i$ fall inside the unit circle on the z-plane, i.e.

$$a^2 + b^2 < 1. \qquad (5.14)$$

Many equivalent theorems in different forms may be obtained by associating Eq.(5.13) with Eq.(5.14). This is intentionally skipped here. The interested reader is referred to the large literature of discrete-time control.

Using the above design method, not only can the stability but also the transient performance of the feedback scheduling system be determined by the locations of the closed loop poles on z-plane. In other words, different but predictable feedback scheduling performance can be achieved through choosing different desired closed loop poles, i.e. $a \pm b_i$. For the characteristics of system transient behaviours with different poles, please consult textbooks on discrete-time control theory.

It is noteworthy that the above design and analysis are intended for the feedback scheduling loop, which differs from the original control loops within the embedded control system. Though both of them are constructed on the basis of feedback control theory and resemble each other in form, they should never be confused in any applications of feedback scheduling to control systems.

Given below is a simple example to demonstrate briefly how to calculate the feedback scheduler coefficients.

Example 5.1 It is known that $K_\lambda = 1.5$ in Eq.(5.9). Desired closed loop poles are 0.3 ± 0.1i. Determine coefficients K_P and K_I of the corresponding feedback scheduler.

Solution. By substituting $K_\lambda = 1.5$, $a = 0.3$, and $b = 0.1$ into Eq.(5.13), it is obtained that:

$$\begin{cases} 1.5K_P + 1.5K_I - 2 = 0.6 \\ 1 - 1.5K_P = 0.1 \end{cases}.$$

Solving the above equation group gives the result: $K_P = 0.6$, $K_I = 1.13$.

The work flow of the feedback scheduling system designed above is as follows. During each invocation interval, the system monitors the actual CPU utilization. When the feedback scheduler is activated at the j-th time instant, it first samples current CPU utilization $U(j)$, then compares it with the desired level. Based on the difference, $\Delta\beta(j)$ is calculated using the PI algorithm. Once α is computed by

$$\alpha(J) = \frac{1}{\beta(j)} = \frac{1}{\beta(j-1) + \Delta\beta(j)},$$

it will be multiplied by the gain scheduling component $1/\hat{\omega}(j)$. After that, the processor alters its supply voltage and clock frequency accordingly. The pseudo code of this scheme is given in Fig.5.2.

Remark 5.3 Real processors can only support a limited number of allowable voltage levels. Therefore, the assumption of arbitrary continuous voltage scaling is by no means true. Taking this into account, some modifications on the above algorithm are mandatory before applying it to real processors with discrete voltage levels. The relevant method will be explored in Chapter 6.

Remark 5.4 With the goal of reducing CPU energy expenditure, this chapter indirectly adapts the execution time of control tasks through dynamically adjusting the supply voltage and operating frequency of the processor. Task execution time are prolonged because of the decrease of CPU speed. The CPU will not always operate at its full speed. A direct consequence is that the delays in control loops increase. Moreover, the dynamic changes in CPU speed will further aggravate the variability of control delays. As a result, the QoC of the system may be

```
//U: Actual CPU utilization
//w: Estimated CPU workload
//α: Normalized CPU speed
Energy Aware Feedback Scheduling {
    Measure U and w;
    //Calculate control input
    Compute ΔμΔU_R−U;
    ComputeΔ β (w.r.t ΔU) using PI algorithm ;
    Compute α based on Δ β (and β );
    Rescale α with 1/w;
    //Reassign CPU speed
    IF α_min=α=1
        Assign CPU speed at α;
    ELSE IF α>1
        Assign CPU speed at 1;
    ELSE
        Assign CPU speed at α_min;
    END
}
```

Fig. 5.2 Pseudo code for Energy-Aware Feedback Scheduling

deteriorated. To reduce this effect, some methods may be employed in the control loops to compensate for the time-varying delays. In the control community, there exist a large number of these methods, see Chapter 2. Since our focus is not on controller design in the control loops, delay and its jitter will not be compensated for. Instead, their effects on QoC will be examined by simulation experiments.

5.4 Simulation Experiments

5.4.1 Setup Overview

Consider an embedded control system composed of three independent control loops. All controlled processes are inverted pendulums with the same linearized model given by:

$$\begin{aligned} \dot{\boldsymbol{x}} &= \begin{bmatrix} 0 & 1 \\ 100 & 0 \end{bmatrix} \boldsymbol{x} + \begin{bmatrix} 0 \\ 100 \end{bmatrix} \boldsymbol{u} + \boldsymbol{v}(t) \\ \boldsymbol{y} &= [1 \quad 0] \boldsymbol{x} + \boldsymbol{e}(t) \end{aligned} , \tag{5.15}$$

where $\boldsymbol{v}$ and $\boldsymbol{e}$ are sequences of white Gaussian noise with zero mean, and their variances are 0.1 and 0.0001, respectively.

The sampling periods of control loops are given by $h = \{20, 25, 30\}$ ms. All controllers (in the original control loops) are well-designed using LQG control algorithm, where the optimization objective function is:

$$J = \int_0^{\infty} (y^2 + 0.01u^2)\,dt. \tag{5.16}$$

In each run of the simulations, the following accumulative control cost for each control loop is recorded:

$$J_i(t) = \int_0^{t} [y_i^2(\tau) + 0.01u_i^2(\tau)]\,d\tau. \tag{5.17}$$

Intuitively, the larger the value of J the worse the QoC.

In this section four different system design methods are compared, i.e. three most representative traditional methods and the proposed approach.

(1) DVS-0: The processor always operates at its full speed, i.e. there is no DVS scheme.

(2) DVS-1: Traditional DVS scheme based on the tasks' WCET. Because hard real-time is guaranteed in this context, the desired CPU utilization level is set to 100% to make full use of the CPU resource. Accordingly, $\alpha = \sum_{i=1}^{3} \frac{WCET_i}{h_i}$.

(3) DVS-2: Traditional DVS scheme based on the estimated execution time of tasks. It holds that $\alpha = \sum_{i=1}^{3} \frac{\hat{c}_{i,\text{nom}}}{h_i}$.

(4) EAFS: The approach proposed in this chapter. Some related parameters are set as follows: $U_R = 95\%$, $T_{FS} = 100$ ms, $K_P = 0.6$, $K_I = 1.13$ (obtained in Example 5.1).

The minimum allowable scaling factor $\alpha_{\min}$ is set to 0.1, quite a small value, such that its effect on feedback scheduling is disregarded in this chapter, and will be studied in the next chapter. The changing values of the execution time factor λ are given in Table 5.1. The estimated nominal execution time of each control task is $\hat{c}_{i,\text{nom}} = 4$ ms ($i = 1, 2, 3$). Accordingly $WCET_i = 6$ ms.

Table 5.1 Values of execution time factor

Time [s]	0 – 3	3 – 6	6 – 9	9 – 12
λ	0.8	1.0	0.5	1.5

5.4.2 Results and Analysis

Since the objective of Energy-Aware Feedback Scheduling is to save energy while preserving QoC guarantees, it is intuitive that there are two main aspects of the system performance, i.e. energy consumption and QoC. Therefore, the simulation results are analyzed below from these two perspectives respectively.

Energy Consumption

The normalized CPU energy consumption under different schemes is shown in

Fig. 5.3. Since the energy consumption calculated here is a normalized value, it will be given in the form of percentage hereafter.

With the first scheme DVS-0, the processor always operates at the highest possible voltage level, i.e. $\alpha \equiv 1$. Therefore the normalized energy consumption $E(\alpha) \equiv 100\%$. It is clear that the energy consumption is the maximum and there is no capability of saving energy in this case. Under the scheme of DVS-1, the $WCET_i$ and the sampling periods are fixed and in consequence $\alpha \equiv 0.74$, $E(\alpha) \equiv 54.8\%$. The *normalized energy saving*, which is defined as $100\% - E(\alpha)$, is 45.2%. Similarly, when the third scheme DVS-2 is employed, $E(\alpha) \equiv 0.49^2 \times 100\% \equiv 24.0\%$ because $\hat{c}_{i,\,\mathrm{nom}}$ remains constant during runtime. In contrast, CPU energy consumption varies with λ under EAFS. The following characteristics can be observed from Fig.5.3.

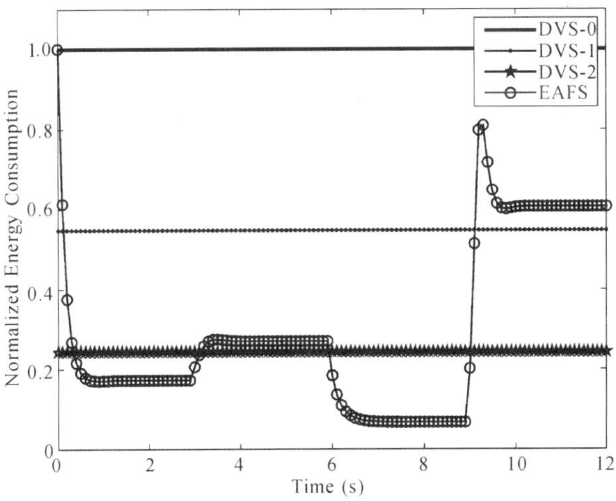

Fig. 5.3 Normalized CPU energy consumption

When the execution time factor λ remains fixed, energy consumption will settle down to a specific level after a short transient process.

(1) The resulting energy consumption in steady states changes with different λ values.

(2) EAFS is capable of dealing with different kinds of unpredictable workload variations such as abrupt increase and decrease in the λ values.

Throughout the simulation the average energy consumption in the case of EAFS comes to 29.7%, which is a little higher than that under DVS-2, but 25.1% lower than that under DVS-1.

From the viewpoint of energy saving, CPU idle time means waste of both computing resource and energy, and hence it should be minimized to maximize the utilization of available CPU resources. According to this observation, the reasons behind the above results can be explained by the requested CPU utilization of all tasks. Note that the requested CPU utilization is not necessarily equal to the actual

CPU utilization, because the actual CPU utilization can never be higher than 100% whereas the requested CPU utilization might be.

As shown in Fig.5.4, the (requested) CPU utilization under DVS-0 is always the lowest. For instance, the CPU utilization is as low as 25% in the time interval t = 6 – 8 s, which implies serious resource waste. In a like manner, the CPU utilization under DVS-1 is also lower than that under both DVS-2 and EAFS, and does not exceed the schedulability bound of the EDF algorithm, i.e. 100%. Therefore, the performance of DVS-1 in saving energy is worse than those of DVS-2 and EAFS. Under DVS-2, the requested CPU utilization changes with λ. When λ is relatively small, DVS-2 results in significant resource waste. For instance, the CPU utilization under DVS-2 is only 50% when $\lambda = 0.5$. In contrast to the dramatic fluctuations of requested CPU utilization under DVS-2, the requested CPU utilization under EAFS is quite steady. Except for the transient processes, the CPU utilization keeps at the desired high level (i.e. 95%) most of the time. This indicates that the CPU time is almost fully used in the case of EAFS.

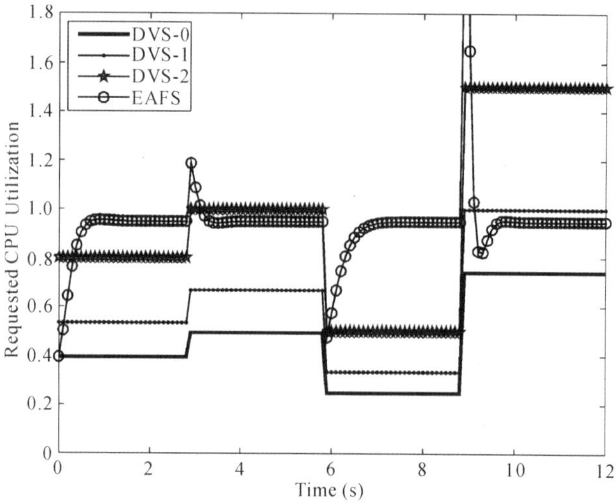

Fig. 5.4 Requested CPU utilization

It is possible that the system is temporarily overloaded when EAFS is used. For example, in the short time interval after $t = 9$ s, the abrupt increase of execution time factor from 0.5 to 1.5 incurs the fact that the requested CPU utilization is temporarily higher than the schedulability bound. In this situation some deadlines are missed, which would negatively impact the QoC. As a future work, methods will be developed for overload handling.

The CPU schedule clearly illustrates the use of CPU resources under different schemes, see Fig.5.5. It can be seen from Fig.5.5(a) that much of the CPU time is wasted when DVS-0 is employed. There is a large amount of time at which no task is running. Fig.5.5(b) indicates that the CPU idle time under DVS-1 reduces in comparison with the case of DVS-0, but still exists. It is also possible that DVS-2 would waste significant CPU time, e.g. in the time interval $t = 8.5 - 9$ s, as shown in Fig.5.5(c). Under EAFS Fig.5.5(d), the CPU idle time is relatively small. For instance, the CPU utilization is as high as 95% in the time interval $t = 8.5 - 9$ s. This explains in a sense why EAFS is so effective in reducing energy consumption.

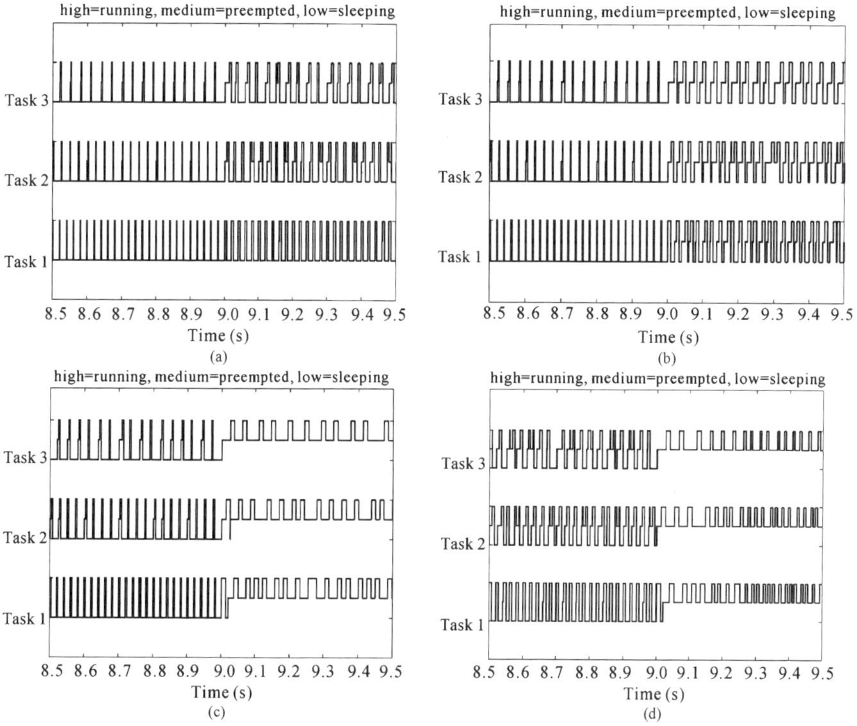

Fig. 5. 5 CPU schedule under (a) DVS-0, (b) DVS-1, (c) DVS-2 and (d) EAFS

QoC

In the following the control performance under different schemes is examined individually first, and then compared.

When the first scheme DVS-0 is applied, the actual execution time of control tasks are always the shortest due to the highest CPU speed. This leads to the smallest control delays. Fig.5.6 depicts the curve of the system output, i.e. the angle of pendulum from the vertical direction, of each loop, while Fig.5.7 shows the control input of each loop.

It is evident that in this case the QoC of all three control loops is quite good all the time. Note that the difference in control performance between the control loops primarily derives from their different choices of sampling periods and the stochastic property of the noises in the systems. However, this difference will not affect the comparison of the different schemes.

In the case of DVS-1, the pendulum angle and the control input of each control loop are given in Figs.5.8 and 5.9, respectively. The control performance is also good.

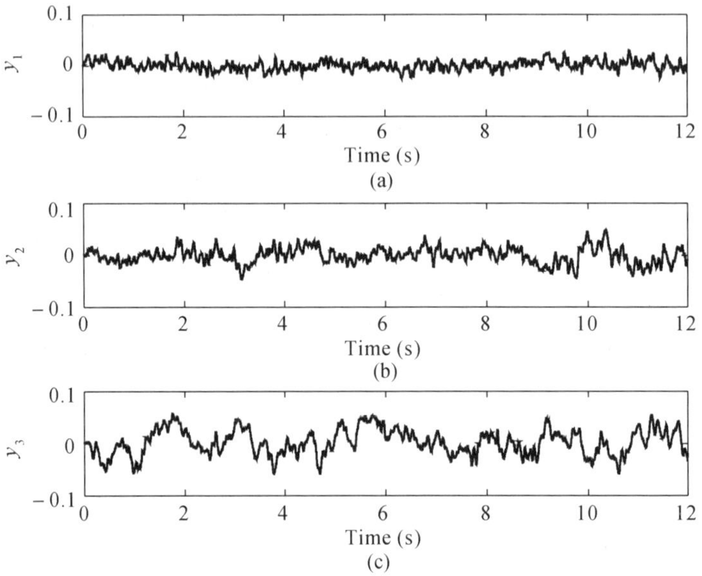

Fig. 5.6 System output under DVS-0
(a) Inverted Pendulum 1; (b) Inverted Pendulum 2; (c) Inverted Pendulum 3

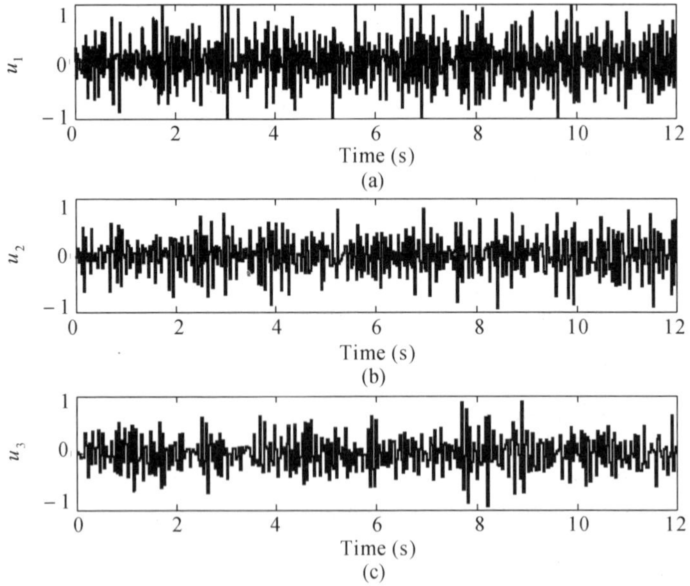

Fig. 5.7 Control input under DVS-0
(a) Inverted Pendulum 1; (b) Inverted Pendulum 2; (c) Inverted Pendulum 3

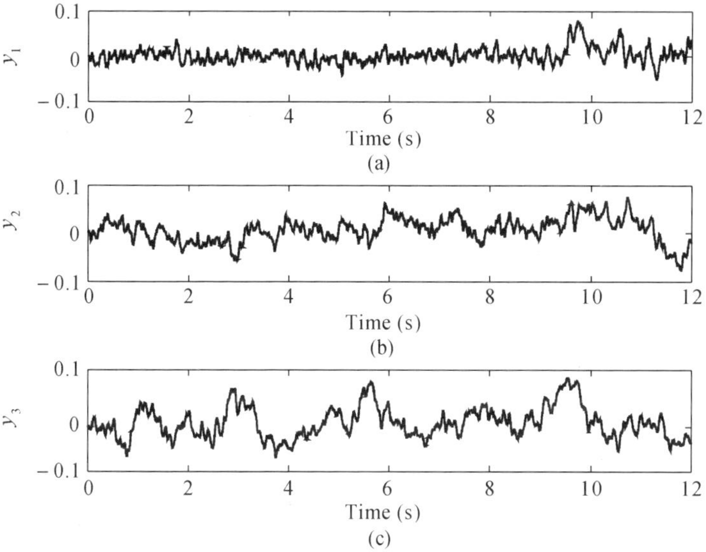

Fig. 5.8 System output under DVS-1
(a) Inverted Pendulum 1; (b) Inverted Pendulum 2; (c) Inverted Pendulum 3

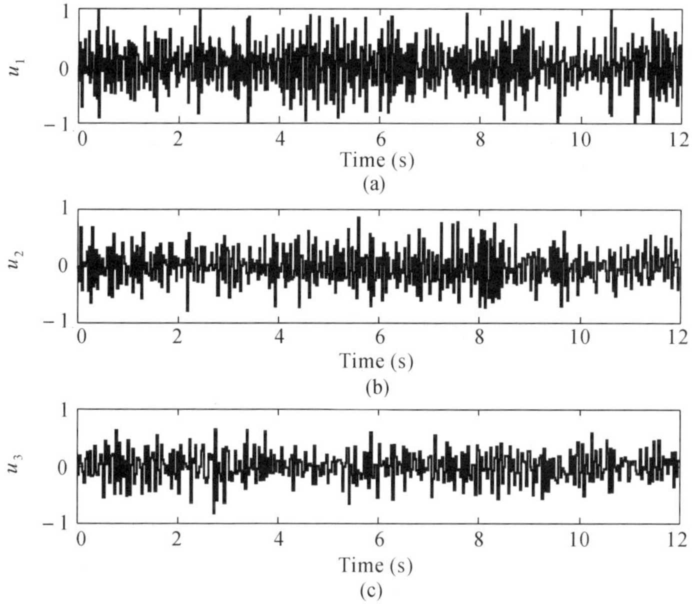

Fig. 5.9 Control input under DVS-1
(a) Inverted Pendulum 1; (b) Inverted Pendulum 2; (c) Inverted Pendulum 3

Figs.5.10 and 5.11 show the control performance under DVS-2. The QoC is good during the time $t < 9$ s. But the system goes unstable after $t = 9$ s. It can be seen

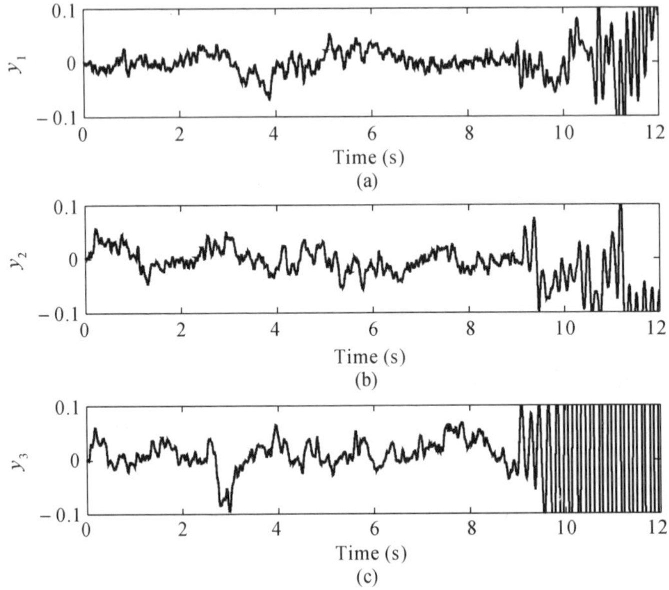

Fig. 5. 10 System output under DVS-2
(a) Inverted Pendulum 1; (b) Inverted Pendulum 2; (c) Inverted Pendulum 3

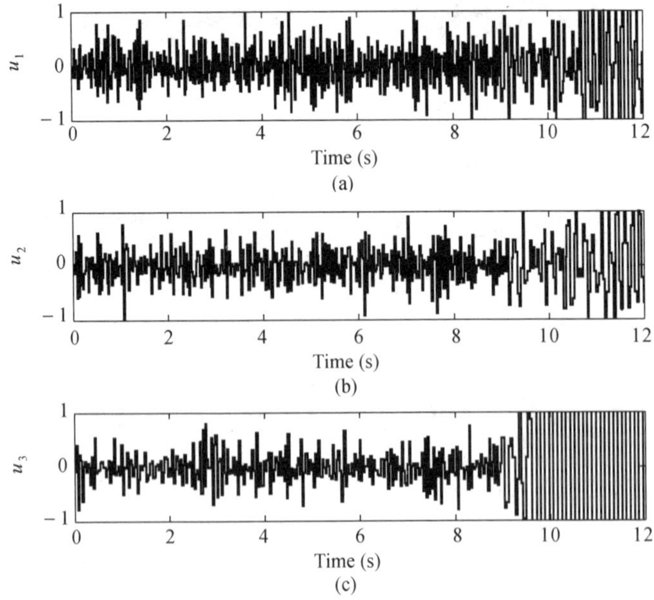

Fig. 5. 11 Control input under DVS-2
(a) Inverted Pendulum 1; (b) Inverted Pendulum 2; (c) Inverted Pendulum 3

from Fig.5.4 that the requested CPU utilization increases up to 150% when $t > 9$ s, which is rather higher than the schedulability bound of the system. As a consequence, the system is severely overloaded. Also, the CPU schedule on Fig.5.5(c) shows that all control tasks are continuously preempted after the time instant $t = 9$ s, which incurs frequent deadline misses. This is why the system cannot maintain stability.

The system output and control input under EAFS are depicted in Figs.5.12 and 5.13, respectively. It is clear that the QoC of each control loop is quite good throughout the simulation.

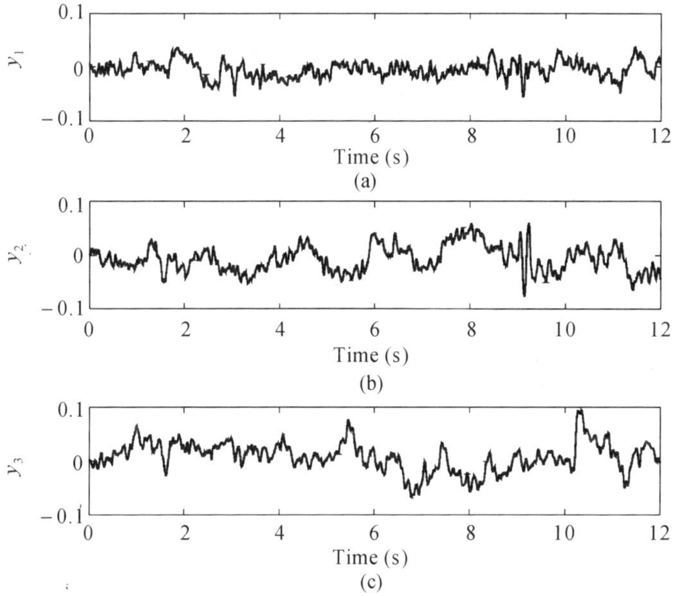

Fig. 5.12 System output under EAFS
(a) Inverted Pendulum 1; (b) Inverted Pendulum 2; (c) Inverted Pendulum 3

Fig.5.14 gives the sum of accumulative control costs of the three loops, i.e. $\sum J_i$. Obviously, all schemes except for DVS-2 achieve satisfactory control performances. Just as expected, the QoC under DVS-0 is the best. Once the DVS technique is introduced under DVS-1, the control delays increase slightly, which causes minor degradation in control performance. However, the overall control performance is still comparably good. The performance of EAFS in terms of QoC is almost identical with that of DVS-1.

The following conclusions can be drawn from the above results:

(1) In systems where the task execution time are unpredictable and time-varying, the proposed EAFS approach is capable of not only preserving good control performance, but also reducing remarkably the CPU energy consumption.

(2) Compared with WCET-based or estimated execution time based traditional DVS schemes, EAFS yields much better overall performance.

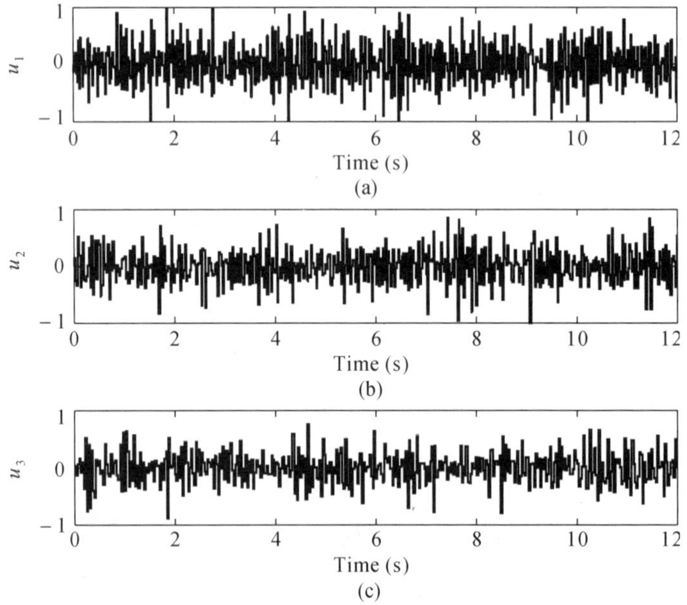

Fig. 5.13 Control input under EAFS
(a) Inverted Pendulum 1; (b) Inverted Pendulum 2; (c) Inverted Pendulum 3

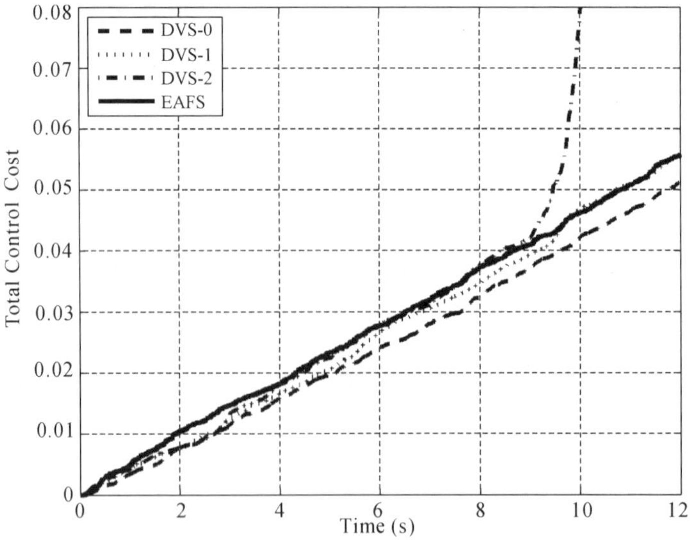

Fig. 5.14 Total control cost of the system

5.5 Summary

This chapter is concerned with energy management in embedded Real-Time Control Systems. Within the framework of feedback scheduling, an energy-aware feedback control scheduling scheme that exploits DVS technology has been presented. A mathematical model has been deduced for the DVS system, and then a control-theoretic feedback scheduler design and analysis method has also been proposed. One major feature of the proposed approach is the integration of real-time control and energy-aware computing.

It has been demonstrated by simulation experiments that the proposed approach performs quite well with respect to both energy saving and QoC guaranteeing in unpredictable environments. In the simulations conducted in this chapter, the proposed approach achieves on average 25.1% additional energy consumption reduction in comparison with the WCET-based scheme. While compared to the traditional scheme based on estimated execution time, the proposed approach performs much better in preserving system schedulability, and consequently provides better QoC.

References

[AST97] K.J. Åström, B. Wittenmark. *Computer Controlled Systems: Theory and Design, 3rd Edition*. Prentice Hall, 1997.

[AYD04] H. Aydin, R. Melhem, D. Mosse, P. Mejia-Alvarez. Power-Aware Scheduling for Periodic Real-Time Tasks. *IEEE Trans. Computers*, Vol. 53, No.5, pp.584 – 600, 2004.

[CER00] A. Cervin, J. Eker. Feedback Scheduling of Control Tasks. *Proc. 39th IEEE CDC*, pp.4871 – 4876, 2000.

[GUT97] V. Gutnik, A.P. Chandrakasan. Embedded Power Supply for Low-Power DSP. *IEEE Trans. VLSI Systems*, Vol.5, No.4, pp.425 – 435, 1997.

[KER05] M. Kersten. Dynamic Voltage Scaling and its Scheduling Implications. *Research Seminar on Energy-Awareness*, Dept. of Computer Sci., Univ. of Helsinki, http://www.cs.helsinki.fi/u/kraatika/Courses/sem05a-EnergyAwareness.html. 2005.

[KWO05] W.C. Kwon, T. Kim. Optimal Voltage Allocation Techniques for Dynamically Variable Voltage Processors. *ACM Transactions on Embedded Computing Systems*, Vol. 4, No.1, pp.211 – 230, 2005.

[LIN04] C. Lin, P. Marti, S.A. Brandt, S. Banachowski, M. Velasco, J.M. Fuertes. Improving Control Performance Using Discrete Quality of Service Levels in a Real-Time System. *10th IEEE Real-Time and Embedded Technology and Applications Symposium*, *Work in Progress Proc.*, 2004.

[LU02] C. Lu, J.A. Stankovic, G. Tao, S.H. Son. Feedback Control Real-Time Scheduling: Framework, Modeling, and Algorithms. *Real-Time Systems*, Vol.23, No.1/2, pp.85 – 126, 2002.

[MEJ04] P. Mejia-Alvarez, E. Levner, D. Mosse. Adaptive Scheduling Server for Power-Aware Real-Time Tasks. *ACM Trans. on Embedded Computing Systems*, Vol.3, No.2, pp.284 – 306, 2004.

[PIL01] P. Pillai, K.G. Shin. Real-Time Dynamic Voltage Scaling for Low Power Embedded Operating Systems. *Proc. 18th ACM Symposium on Operating Systems Principles*, Banff, Alberta, Canada, pp.89 – 102, 2001.

[SIM05] D. Simon, D. Robert, O. Sename. Robust Control/Scheduling Co-design: Application to Robot Control. *Proc. IEEE RTAS*, California, USA, 2005.

[SIN01] A. Sinha, A.P. Chandrakasan. Energy Efficient Real-Time Scheduling. *Proc. Int. Conf. Computer Aided Design*, pp.458 – 463, 2001.

[SOR05] A. Soria-Lopez P. Mejia-Alvarez, J. Cornejo. Feedback Scheduling of Power-Aware Soft Real-Time Tasks. *Proc. 6th Mexican Int. Conf. on Computer Science (ENC'05)*, pp.266 – 273, 2005.

[UNS03] O.S. Unsal, I. Koren. System-level Power-aware Design Techniques in Real-Time Systems. *Proceedings of the IEEE*, Vol.91, No.7, pp. 1055 – 1069, 2003.

[WEI94] M. Weiser, B. Welch, A. Demers, S. Shenker. Scheduling for Reduced CPU Energy. *First USENIX Symposium on Operating Systems Design and Implementation*, Montreal, Canada, pp.13 – 23, 1994.

[XIA05a] F. Xia, X.H. Dai, X.D. Wang, Y.X. Sun. Feedback Scheduling of Real-Time Control Tasks in Power-Aware Embedded Systems. *Proc. 2nd Int. Conf. on Embedded Software and Systems*, Xi'an, China, pp.513 – 518, 2005.

[XIA05b] F. Xia, Y.X. Sun. NN-based Iterative Learning Control under Resource Constraints: A Feedback Scheduling Approach. *Lecture Notes in Computer Science*, Vol.3498, pp.1 – 6, 2005.

[XIA05c] F. Xia, Y.X. Sun. Anytime Iterative Optimal Control Using Fuzzy Feedback Scheduler. *Lecture Notes in Artificial Intelligence*, Vol. 3682, pp.350 – 356, 2005.

[XIA07] F. Xia, Y.C. Tian, Y.X. Sun, J.X. Dong. Control-Theoretic Dynamic Voltage Scaling for Embedded Controllers, Submitted, 2007.

[ZHA06a] W.H. Zhao, F. Xia. Dynamic Voltage Scaling with Asynchronous Period Adjustment for Embedded Controllers. *Dynamics of Continuous, Discrete and Impulsive Systems-Series B. Applications and Algorithms, Special Issue on ICSCA'06*, pp.514 – 519, Watam Press, 2006.

[ZHA06b] W.H. Zhao, F. Xia. An Efficient Approach to Energy Saving in Microcontrollers. *Lecture Notes in Computer Science*, Vol. 4186, pp. 595 – 601, 2006.

[ZHU05] Y. Zhu. Dynamic Voltage Scaling with Feedback EDF Scheduling for Real-Time Embedded Systems. Ph.D Thesis, North Carolina State University, 2005.

Chapter 6

Enhanced Energy-Aware Feedback Scheduling

Energy-Aware Feedback Scheduling(EAFS) presented in the previous chapter offers a promising approach to reduce the energy consumption of embedded controllers. However, the DVS technique that traditionally exploits fixed timing constraints poses an inherent limitation on its capability in energy saving.

Taking advantage of the basic framework of Energy-Aware Feedback Scheduling (EAFS), this chapter aims to achieve further energy consumption reduction while not jeopardizing the QoC. After the dynamic behaviour of control systems with variable sampling periods is examined, an Enhanced Energy-Aware Feedback Scheduling (EEAFS) scheme exploiting the methodology of graceful degradation is presented. In addition to the adjustment of CPU speed using the DVS technology, the sampling period of each control loop is adapted to its actual control performance, thus exploring flexible timing constraints on control tasks. Since the sampling period is enlarged as much as possible given that the QoC is guaranteed, much less energy is consumed. Simulation experiments demonstrate the merits of EEAFS under different scenarios. Compared with the optimal pure DVS scheme, it saves remarkably more energy, and provides comparable control performance at the same time.

6.1 Introduction

It has been demonstrated in the previous chapter in which integrating feedback scheduling with Dynamic Voltage Scaling is an effective way for managing simultaneously QoC and energy consumption in embedded Real-Time Control Systems. DVS provides an enabling technology for feedback schedulers to manipulate the tasks' execution times, and is essential to the application of feedback scheduling in dynamic energy management.

In dynamic environments featuring workload fluctuations, the Energy-Aware Feedback Scheduling scheme (see Chapter 5) uses the DVS technique to alter CPU supply voltage and clock frequency according to current system workload. The basic idea is to maintain the CPU utilization at a desired level. The objectives are twofold: 1) to reduce CPU idle time as much as possible so that the energy consumption is minimized; 2) to respect the schedulability constraint, thus guaranteeing required QoC. By employing this mechanism, the energy consumption of the processor can be effectively reduced, along with guaranteed control performance.

In the area of DVS, existing real-time DVS algorithms rarely take into account the performance of the target applications other than real-time guarantees when determining the voltage level of the processor. Though a large body of DVS work exists for real-time applications, state-of-the-art DVS algorithms, e.g. AVG [PER98], Cycle-conserving and Look-ahead [PIL01], DR-OTE and AGR [AYD04], usually rely on fixed timing constraints of real-time tasks. They typically derive the processor speed that provides timeliness guarantees during runtime according to pre-specified periods/deadlines of the task set, and the timing attributes will never be intentionally changed in favour of energy savings, e. g. in response to the actual application requirements.

In practice, however, the amount of computing resource that an application demands may vary over time. It is true in many circumstances. One representative example is control systems. In a multitasking embedded control system, the controlled process may be disturbed by diverse perturbations, e. g. a step input change, and hence experiences transient processes, in spite of the successful schedule of the whole control task set. From the control perspective, smaller sampling periods are beneficial to the rapid recovery of steady states. Consequently the negative effect of perturbations will be alleviated, and the QoC will then be improved. When the system is in a steady state, however, an unnecessarily small sampling period implies waste of resources (e. g. CPU time and energy). The sampling period may be enlarged to some extent without significantly degrading the control performance [LIN03, MAR02]. This feature of real-time control applications makes it possible to dynamically allocate CPU resource to each control task according to their real demands.

Like the previous chapter, this chapter is also about energy consumption management in multitasking embedded control systems. The overall objective is still to reduce the CPU energy consumption and to preserve QoC guarantees simultaneously. Since the problem of unpredictable task execution times has been tackled in Chapter 5, this chapter assumes that accurate execution times of all tasks are available at runtime, though they may possibly vary. Here the focus is put on achieving further energy saving within the framework of Energy-Aware Feedback Scheduling by exploiting graceful degradation in control applications.

This chapter presents an EEAFS scheme [XIA06, XIA07a, XIA07b] that takes advantage of application adaptation. It is designed to achieve further energy consumption reduction over the EAFS scheme, while not jeopardizing control performance. For this purpose, the periods and the execution times of control tasks are adapted simultaneously.

The adjustment of the sampling period is based on current actual control performance of the corresponding control loop. The basic idea is that a control loop with a large control error will be assigned a relatively small sampling period, and vice versa. When the controlled process is experiencing a transient process, i.e. the control error is relatively large, a small sampling period will help improve the control performance. Otherwise, when the controlled process is in a steady state, i.e. the control error is relatively small, a large sampling period will benefit energy saving. In this way, flexible timing constraints are imposed on the control tasks. The computing resources are allocated to control tasks based on their real demands at runtime, which makes it possible to further reduce energy consumption. In EEAFS the task execution times are adjusted dynamically using the DVS technique, which is similar to Chapter 5.

Based on flexible timing constraints of control tasks, EEAFS utilizes the DVS technique to dynamically alter the CPU speed in a way that the energy consumption is further reduced under the constraint on system schedulability. Extensive simulation experiments for different scenarios are carried out, highlighting the advantages of the proposed approach. In particular, the performance of EEAFS and the optimal pure DVS scheme is compared. The results argue that EEAFS can dramatically reduce CPU energy consumption while providing comparable control performance.

As mentioned in Chapter 5, the research direction of DVS in embedded controllers is nearly unexplored. One notable work related to EEAFS is [LEE04], where the authors propose to alter the sampling periods according to current states of the controlled processes, with the goal of saving more energy. In particular, the sampling period of a control loop is simply switched between two preset values based on whether or not the relevant controlled process is experiencing a transient process. This is quite preliminary and intuitive. On the contrary, the EEAFS scheme employs specific algorithms to continuously adapt the sampling periods within allowable ranges. In comparison, the EEAFS scheme is more widely applicable and systematic in that it establishes the mapping from control application requirements to resource allocation. Moreover, the concept of flexible timing constraints is entirely exploited in this scheme. EEAFS is more flexible and in consequence may be more effective for tradeoffs between QoC and energy consumption. In addition, the DVS technique is combined with feedback scheduling in this chapter, which is not the case in [LEE04].

The rest of this chapter is structured as follows. In Section 6.2 the system model considered in this chapter is described. An optimal pure DVS scheme is presented, which serves as both a building block of EEAFS and the baseline for comparison. Section 6.3 examines the impact of sampling period on energy consumption and QoC. The motivation of this chapter is illustrated by means of case studies. Section 6.4 details the proposed EEAFS scheme. Described first is the architecture of EEAFS. Two alternative feedback scheduling algorithms are then given, followed by the theoretical performance analysis of EEAFS. To reinforce the practicability of this approach, the problem of how to deal with real processors with discrete voltage levels is addressed. In Section 6.5, several sets of simulation experiments under different scenarios are carried out assessing the performance of the proposed approach. Finally, this chapter is concluded in Section 6.6.

6.2 Optimal Pure Dynamic Voltage Scaling

Consider an energy-limited multitasking embedded controller, which is responsible for executing N—independent control tasks. Each control task corresponds to a physical process. The processor supports the DVS technique. The supply voltage and operating frequency (processor speed) can be modified with a scaling factor $\alpha \in [\alpha_{\min}, 1]$. As in Chapter 5, α is also used hereafter to denote the (normalized) CPU speed. By default, α can take on an arbitrary value in the range of $[\alpha_{\min}, 1]$, i.e. the CPU speed can change continuously. The cases in which the processor only supports limited allowable voltage/speed levels will be handled in Section 6.4.4.

The timing attributes of a control task i are described below:

(1) h_i: period, which is equal to the sampling period of the control loop, with a nominal (initial) value of $h_{i,0}$.

(2) $c_{i,\text{nom}}$: execution time at full CPU speed, i.e. when $\alpha = 1$.

(3) c_i: actual execution time when CPU speed is adapted, and it holds that $c_i = c_{i,\text{nom}}/\alpha$.

All the above timing parameters are variable but available at runtime. The definitions of CPU utilization ($U = \sum \frac{c_i}{h_i}$) and workload ($\omega = U \cdot \alpha = \sum \frac{c_{i,\text{nom}}}{h_i}$) remain the same as in Chapter 5.

The system utilizes the EDF algorithm as the underlying scheduling policy. Accordingly, the schedulability condition is:

$$\sum_{i=1}^{N} \frac{c_i}{h_i} \leqslant 1 \iff \omega \leqslant \alpha. \tag{6.1}$$

Just as in Chapter 5, the following two assumptions are made:

(1) It is always held that $\sum_{i=1}^{N} \frac{c_{i,\text{nom}}}{h_{i,0}} \leqslant 1$.

(2) The switching overheads between different voltage levels of the processor are negligible.

The (normalized) energy consumption of the processor is calculated as:

$$E(\alpha) = \alpha^2. \tag{6.2}$$

For the above system, it is not hard to obtain the optimal voltage level α_{opt}, which yields the minimum CPU energy consumption. The following theorem deriving from the work by Sinha and Chandrakasan [SIN01] is quite easy to understand.

Theorem 6.1 For the multitasking embedded control system described above, the processor will consume the minimum energy while meeting the task schedulability constraint if and only if the CPU speed is set to α_{opt} that is given by:

$$\alpha_{opt} = \sum_{i=1}^{N} \frac{c_{i,\,nom}}{h_i} = \omega\,. \tag{6.3}$$

For a pure DVS scheme, it is the best way to scale voltage using Eq.(6.3). Therefore, this scheme is called the optimal pure DVS, denoted opDVS. This chapter constructs EEAFS on the basis of opDVS, and compares their performance.

For simple description, assume that α_{min} is small enough so that $\alpha_{min} \leqslant \omega$, when continuously variable voltage levels are considered. This ensures that the opDVS scheme is always able to set the optimal voltage level. The effect of α_{min} will be studied when dealing with discrete allowable voltage levels.

6.3 Motivating Examples

In this section the motivation of this work is illustrated via case studies. The impact of sampling period on both energy consumption and QoC will be studied respectively.

6.3.1 Energy Consumption with Different Sampling Periods

Examined first is how the sampling periods of control loops, i.e. task periods, impact the energy consumption of the processor. According to Eqs.(6.2) and (6.3), one can easily obtain that

$$E(\alpha) = \alpha_{opt}^2 = \left(\sum_{i=1}^{N} \frac{c_{i,\,nom}}{h_i}\right)^2. \tag{6.4}$$

Given below is an illustrative example.

Example 6. 1 Consider a system composed of two control loops/tasks. It is known that $c_{1,\,min} = 4$, $c_{2,\,min} = 5$, $h_1 \in [10, 20]$, $h_2 \in [10, 30]$. All these parameters are given in the same time unit. When opDVS is employed, the CPU energy consumption as a function of h_1 and h_2 is depicted in Fig.6.1.

It is evident that the CPU energy consumption decreases with increasing sampling periods. A preliminary conclusion is that increasing sampling periods benefits energy saving.

6.3.2 QoC with Variable Sampling Period

As pointed out above, it is possible to reduce energy expenditure by using larger sampling periods. However, according to sampled-data control theory, the increase in sampling period degrades QoC. Therefore, with the traditional framework of fixed timing constraints, the effectiveness of saving energy by increasing sampling periods is often impaired by significant deterioration of the QoC, which is not allowed in most cases.

To address this problem, this chapter attempts to impose flexible timing constraints on control tasks. The following example is used to illustrate the motivation and to demonstrate the feasibility of this idea.

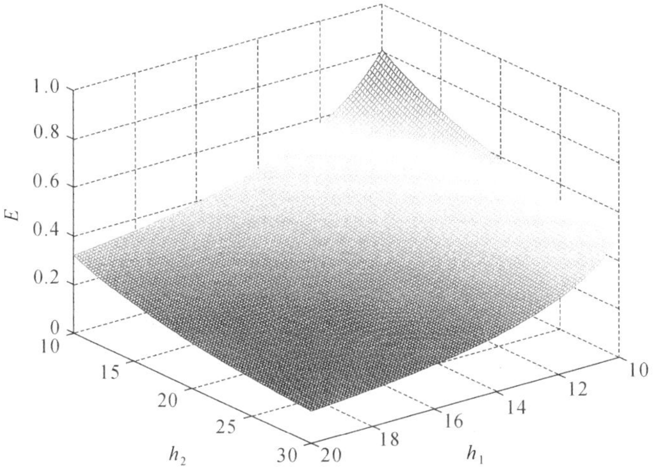

Fig. 6.1 Normalized CPU energy consumption in Example 6.1

Example 6.2 Consider a control loop running on a dedicated processor. The loop setup is the same as the Loop 1 in Section 2.4, with an initial sampling period of 6 ms. Assess the system performance in the following two cases:

(1) Case I: The sampling period remains constant during runtime.

(2) Case II: Increase the sampling period from 6 ms to 12 ms when the system is in steady states.

Fig.6.2 shows the system responses and the task executions under both cases.

From the upper part of Fig.6.2, it can be found that in Example 6.2 the same QoC is achieved under two different cases. According to simulation records, the integrals of absolute error under both cases are the same, i.e. 0.078. In the second case, though the sampling period increases after time instant $t = 0.5$ s, the resulting control performance does not deteriorate at all.

On the other hand, the lower part of Fig.6.2 argues that the increase of sampling period remarkably reduces the consumption of computing resource. In this example, the CPU time that the control task demands decreases to half the original after $t = 0.5$ s, because the sampling period is doubled. According to the result given in Section 6.3.1, it is believed that the CPU energy consumption could be significantly reduced.

It can now be outlined from the above analysis that the sampling period can be properly increased when the relevant control error is relatively small, which benefits energy saving and does not significantly deteriorate control performance. This methodology is referred to as *graceful degradation*, on which this chapter is based. In the EEAFS scheme, variable sampling periods are used in the control loops, so that the CPU energy consumption is further reduced over schemes that exploit pure DVS techniques, while delivering comparable control performance. In this way, the energy efficiency of the system will be improved.

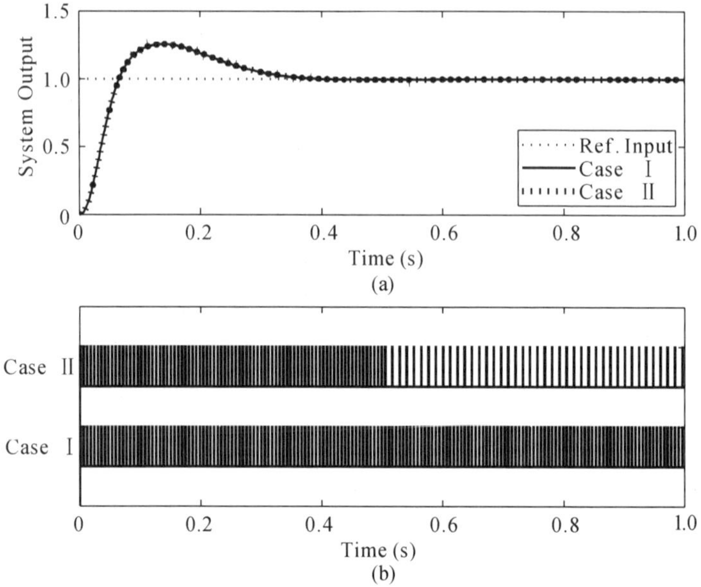

Fig. 6.2 System performance in Example 6.2
(a) System Output; (b) Task Execution

6.4 Enhanced Energy-Aware Feedback Scheduling

Based on the above observation, in this section an Enhanced Energy-Aware Feedback Scheduling (EEAFS) scheme that takes advantage of flexible timing constraints of control tasks is presented. As mentioned previously, the goal is to achieve further energy consumption reduction over pure DVS schemes, without jeopardizing the QoC. In the following, the framework of the proposed approach will be described. Within the same framework, two alternative algorithms are given for period adjustment, followed by a preliminary theoretical performance analysis. A simple yet efficient method for handling the cases of real processors with discrete voltage levels is also suggested.

6.4.1 Framework

The framework of the proposed EEAFS scheme is shown in Fig.6.3.

As in the previous chapter, this chapter also utilizes a time-triggered feedback scheduler, with an invocation interval of T_{FS}. For simplicity, assume that the execution times (at full CPU speed) of all tasks would never change except at the invocation instants of the feedback scheduler. Thus the execution time and the period of each task are deterministic in every invocation interval, which makes the CPU utilization also known. As a consequence, the actual CPU utilization will keep

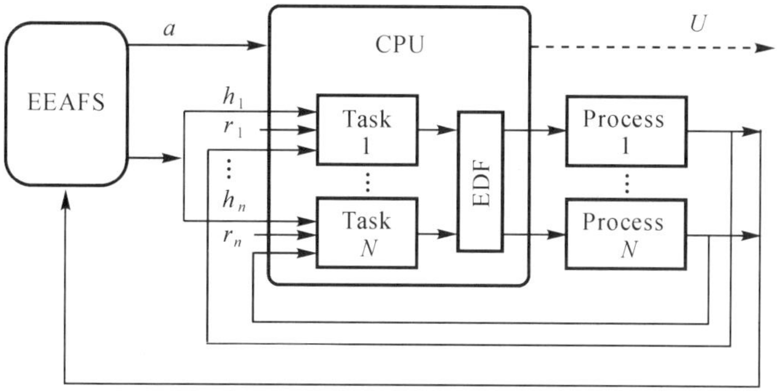

Fig. 6.3 Framework of enhanced Energy-Aware Feedback Scheduling

at the desired level each time the feedback scheduler is executed. In other words, if EEAFS is used, the resulting CPU utilization will always keep at the desired level, 100% for example in this chapter, provided that the time overhead of the feedback scheduler is negligible.

Therefore, in the framework of EEAFS, there is actually no real feedback information about the CPU utilization. This is primarily thanks to the availability of accurate timing attributes of the tasks. The cases where the tasks' timing attributes are uncertain and unpredictable have been attacked in Chapter 5 and therefore will not be considered here.

With EEAFS, the current absolute control error of each control loop will be fed back to the feedback scheduler at every invocation instant. The *absolute control error* e_i is defined as the absolute difference between the reference input r_i and the system output y_i, i.e. $e_i = |r_i - y_i|$.

The feedback scheduler consists of two main parts: period adjustment and voltage scaling. The former is responsible for adjusting the sampling period of each loop on the basis of feedback information about its current control error. The relevant algorithm will be given in the next subsection. The latter exploits the opDVS scheme described in Section 6.2. Its role is of course to scale the voltage/frequency of the processor using the DVS technique. The adoption of opDVS implies setting the desired CPU utilization level to the system schedulability bound, i.e. 100%. This gives rise to the maximum utilization of CPU resources and hence the lowest possible energy expenditure.

In the above description, it is assumed that the sampling periods of all control loops remain constant during the course of every invocation interval. In practice, the following situation may happen. At the j-th invocation instant, the sampling period of a loop, say loop i, is set to the maximum $h_{i,\max}$ because the controlled process is in a steady state. At a certain time instant t ($jT_{FS} < t < (j+1)T_{FS}$), process i is disturbed by a sharp perturbation, e.g. a step input change. Since the sampling period at this time remains to be $h_{i,\max}$, which is too large for the controller to response quickly enough, the corresponding control performance could be significantly degraded.

To deal with this situation, an event-triggered mechanism is introduced for the feedback scheduler in addition to the time-triggered activation. The basic idea behind is that once the absolute change of control error in a control loop exceeds a specific threshold, the feedback scheduler will be invoked immediately. Accordingly, the condition of producing an event that triggers the feedback scheduler is given by:

$$\Delta e_i(t) = | e_i(t) - e_i(jT_{FS}) | > \delta, \tag{6.5}$$

where δ is a design parameter denoting the threshold of Δe_i, and j represents the invocation instant of feedback scheduler. It is possible to choose different δ values for different control loops. For simplicity, this chapter uses the same value $\delta = 0.5$ for all loops in simulation experiments (see Section 6.5).

As Eq.(6.5) argues, the benefits of integrating the event-triggered mechanism are at least twofold.

(1) First, if the absolute control error of a control loop abruptly increases during the course of an invocation interval, the system will invoke the execution of the feedback scheduler and assign a relatively small sampling period to this loop. This benefits the improvement of control performance.

(2) Second, in the case of an abrupt decrease of the absolute control error, the feedback scheduler will also be executed and the relevant sampling period will be enlarged to a certain extent. A consequence is that additional energy consumption reduction can be achieved.

Interesting to note is that when the feedback scheduler is triggered by an event rather than a timer, it will only re-assign the sampling period of the control loop that generates the event, while others remain unchanged. Of course, the CPU speed will still be re-set according to opDVS. One of the major reasons for this operation is to cut down the feedback scheduling overhead.

Thanks to the fact that the EEAFS scheme combines the traditional time-triggered mechanism with an event-triggered mechanism, it will become more cost-effective with improved feedback scheduling performance [ZHA06]. On the other hand, this makes it much easier to choose an appropriate invocation interval T_{FS}. In practice, this choice can be made based on related characteristics of the system, such as the frequency of variations in task execution times, the magnitude and frequency of perturbations. If T_{FS} is a little bigger than necessary, the even-triggered mechanism will compensate for its negative effect, at least in part.

6.4.2 Period Adjustment Algorithms

Now that the voltage scaling module inside the feedback scheduler has been presented in Section 6.2, the key to EEAFS design is the algorithm for sampling period adjustment. As mentioned previously, the basic principle of sampling period adjustment is as follows. When the QoC is good, the corresponding control task will be assigned a large period. Conversely, a small period will be assigned in the case of bad QoC. As long as the schedulability constraint is preserved, the determination of the sampling period for each control loop is only related to its own control performance. That is, the calculation of sampling periods is independent from one loop to another. For simple notation, the subscript i will hereafter be dropped from all variables wherever possible.

A natural prerequisite for online assignment of sampling periods is selecting a proper metric to indicate the instantaneous control performance. Quite a reasonable choice is the absolute control error. It is generally accepted that bigger errors indicate worse QoC. However, if the process output sharply oscillates around the reference input, the absolute control error might still be small sometimes, which obviously does not reflect the real control performance. Taking this into account, the following instantaneous control performance index (denoted *ind*) is defined, which can prevent the QoC feedback information from oscillating too sharply with the control error as well,

$$ind(j) = \lambda \cdot ind(j-1) + (1-\lambda) \cdot e(j) , \tag{6.6}$$

where λ is a forgetting factor. Recall that the invocation interval of the feedback scheduler is different from the sampling periods of control loops.

Before describing how to calculate the sampling period h, its allowable range should be determined. In general, the maximum allowable value of h can be obtained based on the stable condition of the control system. In the control community, as mentioned in Chapter 2, there exist quite a few theories and methods for this purpose. In this monograph, the maximum sampling period h_{max} is instead determined by simulations. In fact, it has been recognized that the system may still be practically stable even if the sampling period is (temporarily) somewhat larger than the theoretical stability bound when the system is in a steady state [ARZ99, GRA04]. This could happen given that the sampling period is sufficiently small when the system is experiencing transient processes. In addition, this further validates the practicability of the employed simulation-based method for determining h_{max}.

As regarding the minimum allowable value of h, it is set for simplicity that $h_{min} = h_0$, i.e. the minimum allowable period is equal to its nominal value. As a matter of fact, it is possible in most cases that a sampling period less than h_0 can be assigned when the control loop encounters a severe perturbation so that the control performance is further improved without violating the system schedulability. This is because some other sampling periods may be increased to cut down their resource demands at system runtime, which makes it possible for the control loop experiencing a transient process to consume more computing resources. However, this will complicate the problem. If sampling periods less than h_0 are allowed, much care should be put on the system schedulability, and the resource allocation should be balanced among control loops according to their demands. Moreover, the improvement of control performance is at the penalty of increasing energy expenditure.

On the other hand, according to $\sum_{i=1}^{N} \frac{c_{i,\mathrm{nom}}}{h_{i,0}} \leqslant 1$, the runtime workload is given by $\omega = \sum_{i=1}^{N} \frac{c_{i,\mathrm{nom}}}{h_i} \leqslant \sum_{i=1}^{N} \frac{c_{i,\mathrm{nom}}}{h_{i,0}} \leqslant 1$, if $h_{min} = h_0$. That is, the feasibility of the voltage scaling module, i.e. opDVS, is then always guaranteed.

Based on the above discussion, the allowable range of sampling periods can be specified accordingly. Illustrated next is how to determine the sampling period dynamically. The formula used here to calculate the new period is as follows:

$$h(j) = \eta(j) \cdot h_{\min} = \eta(j) \cdot h_0, \tag{6.7}$$

where η is the *period scaling factor*. In the following, two alternative algorithms are presented to compute η, an exponential algorithm and a linear algorithm.

Exponential Algorithm

In this case η is computed as:

$$\eta(j) = \begin{cases} \dfrac{h_{\max}}{h_{\min}} & \text{if } ind(j) \leqslant e_{\min} \\ \dfrac{\dfrac{h_{\max}}{h_{\min}} - 1}{e^{-\beta \cdot e_{\min}} - e^{-\beta \cdot e_{\max}}} (e^{-\beta \cdot ind(j)}) - e^{-\beta \cdot e_{\max}} + 1 & \text{if } e_{\min} < ind(j) < e_{\max} \\ 1 & \text{if } ind(j) \geqslant e_{\max} \end{cases}, \tag{6.8}$$

where β is a constant introduced to enhance the effect of the exponential function, $e_{\min}$ and $e_{\max}$ are design parameters that decide the range of the instantaneous control performance index in which sampling period can be adjusted arbitrarily.

According to Eqs.(6.7) and (6.8), the sampling period will be directly set to the minimum, which corresponds to $\eta = 1$, once the performance index $ind(j)$ exceeds a upper threshold $e_{\max}$. The goal of this operation is to improve control performance via quick response to large derivations. In contrast, the period will be the maximum, i.e. $\eta = h_{\max}/h_{\min}$, if $ind(j)$ becomes less than another limit $e_{\min}$, which implies that the system approaches a steady state. The maximum period is set to achieve the largest possible energy saving. In other cases, i.e. when $e_{\min} < ind(j) < e_{\max}$, a period that decreases exponentially with increasing $ind(j)$ will be assigned.

Several example curves of the exponential period rescaling factor function with different β values are depicted in Fig.6.4, where the x-axis is ind and the y-axis is η.

It is apparent that, with the exponential scaling function, the period will rapidly decrease as the performance index (control error) increases. This benefits the improvement of control performance. As β increases, the effect of the exponential function will be enhanced. Since a small η value yields a relatively small sampling period, the algorithm becomes more aggressive with smaller β, from the viewpoint of energy saving.

Linear Algorithm

A more straightforward method to adjust sampling periods is the linear function given below:

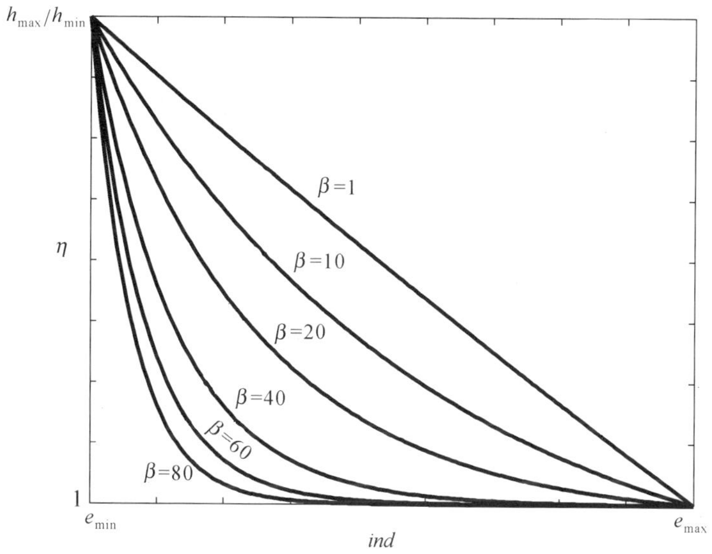

Fig. 6.4 Example curves of exponential period scaling factor function

$$\eta(j)=\begin{cases}\dfrac{h_{max}}{h_{min}} & \text{if } ind(j)\leqslant e_{min}\\ \dfrac{h_{max}}{h_{min}}-\dfrac{(\mathrm{ind}(j)-e_{min})(h_{max}/h_{min}-1)}{e_{max}-e_{min}} & \text{if } e_{min}<ind(j)<e_{max}\\ 1 & \text{if } ind(j)\geqslant e_{max}\end{cases}. \quad (6.9)$$

In the case of either $ind(j) \geqslant e_{max}$ or $ind(j) \leqslant e_{min}$, the computation of this linear function is the same as the exponential function, but they are different when $e_{min} < ind(j) < e_{max}$. In these cases, the linear algorithm adjusts the sampling periods linearly. Its curve is exactly a line that goes across points (e_{min}, h_{max}/h_{min}) and (e_{max}, 1) on the plane. Compared with the exponential function, the linear function is more simple and straightforward, but less flexible.

In both algorithms, it is essential to choose proper values for e_{min} and e_{max}. For the purpose of improving QoC, it is better that the sampling period is set smaller. Therefore, e_{min} and e_{max} should be minimized. However, from the perspective of energy saving, a larger sampling period is better, and in consequence large e_{min} and e_{max} should be chosen. Obviously, improving QoC and reducing energy consumption pose conflicting requirements. The principle of choosing e_{min} and e_{max} is to use as large values as possible given that the QoC is not jeopardized. This largely facilitates further energy saving and hence improves the energy efficiency of the system.

6.4.3 Performance Analysis

The Enhanced Energy-Aware Feedback Scheduling scheme can be described by Fig. 6.5. It can be found that both the execution times and the periods of control

```
//h: Sampling period
//α: Normalized CPU speed
Enhanced Energy Aware Feedback Scheduling {
    IF triggered by timer
        FOR each control loop
          Sample the process output y;
          e←abs(r-y);                    //r: reference input
          Compute ind using Eq.(6.6);
          Compute η using Eq.(6.8) or Eq.(6.9); //η: period rescaling factor
          Compute h using Eq.(6.7);
          // Update task period;
        END
    ELSE (triggered by Loop i)
        Update h_i using Eq.(6.6)–Eq.(6.9);
    END
    Compute α using Eq.(6.3);
    IF voltage levels are discrete
        IF α=α_m
        ELSE
            IF α<α_min Then α← α_min;    //α_min: minimum allowable speed
            IF α_m<α<α_m+1 Then α←α_m+1;
            // Resource reclaiming
            Rescale all sampling periods using Eq.(6.11);
        END
    END
    Assign CPU speed to α;
}
```

Fig. 6.5 Pseudo code for enhanced Energy-Aware Feedback Scheduling

tasks are adapted simultaneously with this scheme. It is the DVS technology that realizes the adjustment of task execution times, while the re-assignment of task period of each loop is based on its actual control performance.

Because the magnitudes of task execution times are highly related to the CPU speed, which also determines the energy expenditure, the adjustment of execution times is associated with the management of CPU energy expenditure. Furthermore, large execution times imply low CPU speeds, which yield low energy consumption. On the other hand, smaller sampling periods deliver better control performance. Under specific resource constraints, i.e. the CPU utilization bound for system schedulability, increasing execution times and decreasing periods conflict with each other, since both cause the requested CPU utilization of control tasks to increase. From this insight, it is the simultaneous adjustment of task execution times and periods that the EEAFS scheme employs to make the compromise between energy consumption and QoC. Recall that the proposed approach enlarges sampling periods on condition that the control performance is not significantly degraded, thereby capable of saving additional energy while providing comparable QoC.

When the EEAFS scheme is applied, the maximum and minimum possible values of the normalized CPU energy consumption are given by the following theorem.

Theorem 6.2 For the multitasking embedded control system described in Section 6.2, if the feedback scheduling depicted in Fig.6.5 is employed, then with given task execution times, the range of normalized CPU energy consumption is $[E_{\min}, E_{\max}]$, where

$$E_{\min} = \left(\sum_{i=1}^{N} \frac{c_{i,\,nom}}{h_{i,\,\max}} \right)^2, \quad E_{\max} = \left(\sum_{i=1}^{N} \frac{c_{i,\,nom}}{h_{i,\,0}} \right)^2. \tag{6.10}$$

The proof of this theorem is quite straightforward. Given below are some brief arguments.

Eq.(6.4) indicates that the processor will consume the minimum of energy if all task periods take on their maximum possible values. According to the algorithms given in Section 6.4.2, the maximum possible value of h_i is $h_{i,\,\max}$. If $\mathrm{ind}(j) \leqslant e_{\min}$ is satisfied simultaneously for all control loops, then the CPU energy consumption reaches the minimum, and holds that

$$E = E_{\min} = \left(\sum_{i=1}^{N} \frac{c_{i,\,\mathrm{nom}}}{h_{i,\,\max}} \right)^2.$$

Similarly, energy consumption reaches its maximum value

$$E = E_{\max} = \left(\sum_{i=1}^{N} \frac{c_{i,\,\mathrm{nom}}}{h_{i,\,0}} \right)^2,$$

when $ind(j) \geqslant e_{\max}$ is satisfied for all loops.

In Example 6.1, for instance, the minimum possible (normalized) energy consumption is $E_{\min} = (4/20 + 5/30)^2 \times 100\% = 13\%$, while the maximum value is $E_{\max} = (4/10 + 5/10)^2 \times 100\% = 81\%$.

Theorem 6.2 gives only the range of all possible instantaneous values of CPU energy expenditure. The actual energy consumption of a practical system is closely related to the states of all controlled processes during runtime. Generally speaking, more heavy perturbations on control loops lead to smaller task periods on average, which result in turn in higher energy expenditure. This will be validated in Section 6.5.

6.4.4 Handling Discrete Voltage Levels

In the previous sections, it is assumed that the CPU voltage can be scaled continuously. However, this can never be true for real processors due to technical reasons. It is only possible for a real processor to support limited variable voltage levels. That is, the allowable values for α are limited. Let $\alpha \in \{\alpha_1, \alpha_2, \cdots, \alpha_m\}$, where $\alpha_1 = \alpha_{\min}$, $\alpha_m = 1$ and $\alpha_m < \alpha_{m+1}$. Moreover, $\alpha_{\min}$ could not be arbitrarily small, which should be accounted for in practical applications.

In this chapter a simple method is employed to handle this situation. After the optimal CPU speed α_{opt} is computed assuming continuous voltage levels, it is then processed as follows:

(1) If $\alpha_{\mathrm{opt}} < \alpha_{\min}$, then set the CPU speed α to $\alpha_{\min}$;

(2) If $\alpha_{\text{opt}} = \alpha_m$, then $\alpha = \alpha_m$;

(3) If $\alpha_m < \alpha_{\text{opt}} < \alpha_{m+1}$, then $\alpha = \alpha_{m+1}$.

Extended with this processing method, the above-described EEAFS algorithm can be applied to real processors with discrete voltage levels.

Not surprisingly, the quantization of allowable voltage levels may possibly cause actual CPU utilization to be lower than 100%. This is primarily because, when $\alpha_m < \alpha_{\text{opt}} < \alpha_{m+1}$, it is imperative to set the CPU speed α to α_{m+1} in order not to violate task schedulability. As a consequence, the actual CPU utilization will be $U_{\text{act}} = \alpha_{\text{opt}}/\alpha_{m+1} < 100\%$. One may have noticed that adding the number of allowable voltage levels M will reduce the impact of discreteness. In addition, the minimum allowable CPU speed $\alpha_{\min}$ may cause unused computing resources as well.

Alternatively, methods could be used to avoid significant resource waste particularly when the number of voltage levels is relatively small. Given below is a simple resource reclaiming mechanism this chapter exploits.

The basic idea behind the resource reclaiming mechanism is to properly decrease the task periods when the CPU utilization is expected to be less than 100% so that the remaining CPU resource resulting from the quantization of voltage levels is also used. One straightforward method to update the sampling periods is given in Eq.(6.11).

$$\tilde{h}_i(j) = \max\left\{h_{i,\min}, \frac{\alpha_{\text{opt}}}{\alpha} h_i(j)\right\}. \tag{6.11}$$

It can be seen that Eq.(6.11) attempts to scale all task periods with $\alpha_{\text{opt}}/\alpha$. In this way, the waste of resource due to discrete voltage levels is minimized, since the free computing resource is used to improve the control performance of each loop. Of course, due to the constraint of $h_{\min}$, a 100% utilization cannot always be guaranteed by Eq.(6.11). To maximize the utilization of the processor, it is possible to further decrease those periods that satisfy $h > h_{\min}$. For the sake of simplicity, this chapter updates task periods according to Eq.(6.11) only once when dealing with processors with discrete voltage levels.

6.5 Performance Evaluation

In this section several sets of simulation experiments are performed to assess the performance of the EEAFS scheme. It will be compared with the optimal pure DVS scheme opDVS whenever possible. Extensive results are presented and analyzed.

Consider an embedded control system consisting of four independent control loops. All controlled processes are selected from [MES06], as shown in Fig.6.6. The controllers are well-designed using the prevalent PID control algorithm, with a continuous-time form given by:

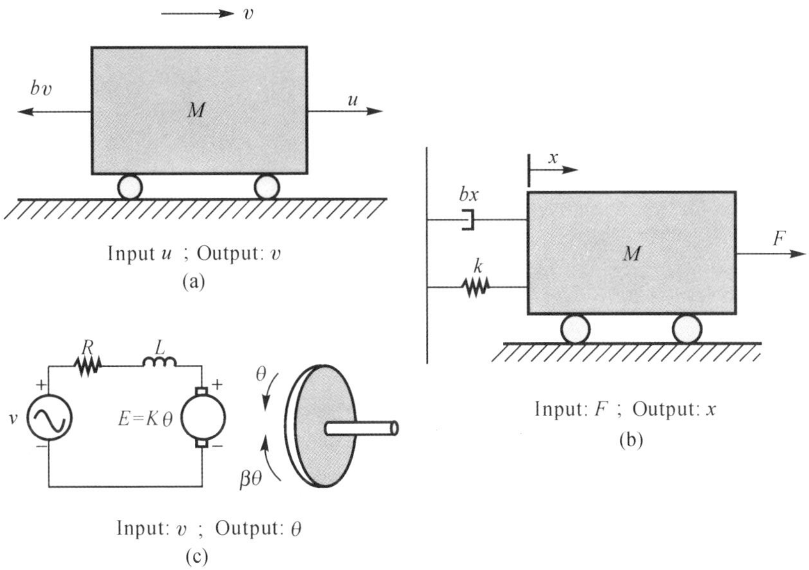

Fig. 6. 6 Controlled processes in control loops
(a) Loop1: Cruise Control; (b) Loop2 and Loop4: Mass, Spring and Damper; (c) Loop3: Motor Speed

$$G_{\mathrm{PID}}(s) = K_{\mathrm{P}} + \frac{K_I}{s} + K_{\mathrm{D}}s\,. \tag{6.12}$$

Digital controllers are designed by discretizing continuous-time controllers, which are expressed as:

$$\begin{aligned} e(k) &= r(k) - y(k) \\ P(k) &= K_{\mathrm{P}}e(k) \\ I(k) &= I(k-1) + K_{\mathrm{I}}h[e(k) + e(k-1)]/2\,. \\ D(k) &= K_{\mathrm{D}}[e(k) - e(k-1)]/h \\ u(k) &= P(k) + I(k) + D(k) \end{aligned} \tag{6.13}$$

The system models of the controlled processes and the corresponding controller parameters are given in Table 6.1, where the time unit of task execution times and periods is in *ms*. For the purpose of simple description, the c_{nom} values of all tasks are fixed in simulations. Because all timing attributes of tasks are precisely known, the proposed scheme will perform in like manner in the case of variable c_{nom} values.

In addition, the following parameters are used in all simulation experiments: $T_{\mathrm{FS}} = 50$ ms, $\lambda = 0.3$, $e_{\max} = 0.2$, $e_{\min} = 0.02$. The perturbations on control loops are step input changes.

Table 6.1 Setup of simulation experiments

	System model	Controller parameters	c_{nom}	$h_0(h_{min})$	h_{max}
Loop 1	$\frac{1}{1000s+50}$	$K_P = 10^4$, $K_I = 400$, $K_D = 0$	2	10	40
Loop 2	$\frac{1}{s^2+10s+20}$	$K_P = 30$, $K_I = 70$, $K_D = 0$	2	7	30
Loop 3	$\frac{1}{0.5s^2+6s+10}$	$K_P = 100$, $K_I = 200$, $K_D = 2$	2	8	30
Loop 4	$\frac{1}{s^2+10s+20}$	$K_P = 200$, $K_I = 350$, $K_D = 3$	2	9	40

To measure the QoC, the Integral of Absolute Error (IAE) is recorded respectively for each loop, i.e. $J_i(t) = \int_0^t e_i(\tau)\,\mathrm{d}\tau$. The total control cost of the system is calculated as $J_{SUM}(t) = \sum_{i=1}^{3} J_i(t)$. Normalized CPU energy consumption is computed using Eq.(6.2).

6.5.1 Experiment Ⅰ: Different Schemes

In this set of simulations, the following schemes are compared:

(1) opDVS: the optimal pure DVS scheme, with fixed task periods;

(2) EEAFS-1: the proposed EEAFS scheme that adopts the exponential period scaling algorithm, with $\beta = 40$;

(3) EEAFS-2: the proposed EEAFS scheme that adopts the linear period scaling algorithm.

The simulation runs as follows. At time $t = 0$, Loops 1 and 2 start running, and Loop 1 is disturbed by a step input change. Loop 2 is perturbed at time $t = 2$ s. Loops 3 and 4 remain off until $t = 4$ s, at which both of them encounter step input changes. At time instant $t = 6$ s, all loops are perturbed simultaneously. The whole run ends at $t = 8$ s.

Energy Consumption

Fig.6.7 shows the normalized CPU energy consumption under the different schemes. As an optimal pure DVS scheme, opDVS is effective in energy consumption reduction, especially when the workload is light, e.g. in the time interval $t = 0 - 4$ s. Compared to opDVS, both EEAFS-1 and EEAFS-2 are able to save much more energy. Throughout the simulations, the average energy consumption under three schemes is 57.9%, 10.9% and 9.7%, respectively. The CPU energy consumption under EEAFS-1 and EEAFS-2 decreases over opDVS by 47.0% and 48.2%, respectively, on average. By instantaneous samples, the reduction of energy consumption under EEAFS-1 and EEAFS-2 versus opDVS reaches up to 86.3%, for example, in the time intervals $t = 5 - 6$ s and $7 - 8$ s. It is also found that the CPU utilization remains at 100% in all simulations. This implies that the CPU time is

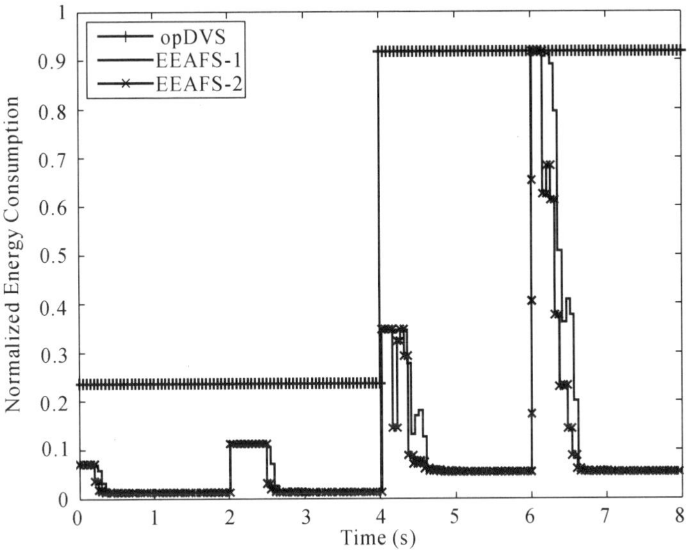

Fig. 6.7 Normalized energy consumptions under different schemes

fully utilized under all three schemes.

The performance difference shown in Fig.6.7 can be explained by the variations of task periods under different schemes. The runtime variation of each task's period is depicted in Fig.6.8. From Fig.6.8, it is not hard to find out that:

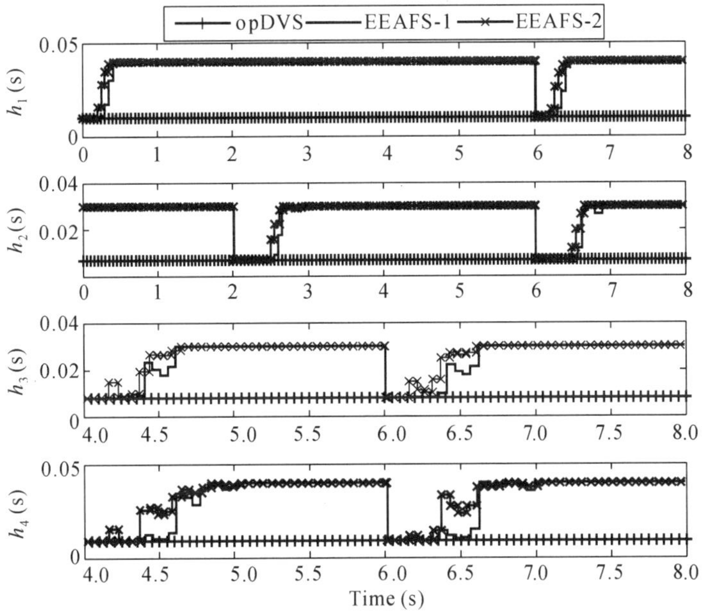

Fig. 6.8 Task periods under different schemes

(1) The sampling periods are the largest under EEAFS-2 almost all the time;

(2) The sampling periods under EEAFS-1 are slightly smaller than but very close to those under EEAFS-2;

(3) The sampling periods under opDVS are always the smallest.

Thanks to that the task periods are enlarged, EEAFS-1 and EEAFS-2 are capable of saving more energy over opDVS. For instance, in time intervals $t = 5-6$ s and $7-8$ s, all task periods are set to the maximum possible values, i.e. $h_i = h_{i,\max}$ when EEAFS is applied. This results in a low energy consumption of 5.5%, which is 86.3% less than that of the opDVS case.

The difference between EEAFS and opDVS in assigning task periods can be clearly reflected by CPU schedule (see Figs.6.9, 6.10 and 6.11). When opDVS is used, all task periods remain changeless regardless of the perturbations on control loops, as shown in Fig.6.9. On the contrary, the sampling period of each loop varies over time in the case of EEAFS (see Figs.6.10 and 6.11). It can be seen from the CPU schedule that the control tasks execute more frequently when the control errors are bigger, which indicates smaller periods. When the system is in a steady state, e.g. in the interval $t = 3.5-4$ s, the task periods are conside-rably larger under EEAFS than under opDVS.

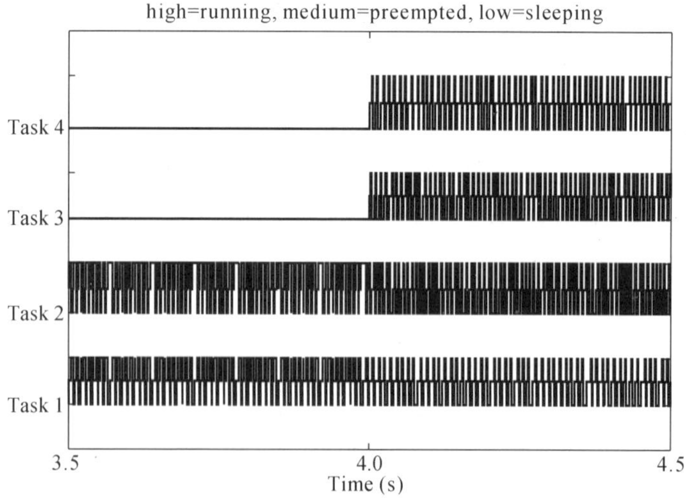

Fig. 6.9 CPU schedule under opDVS

The scaling of voltage/frequency can be examined through observing the variations in task execution times on the CPU schedule. Comparing Figs.6.10 and 6.11 with Fig.6.9, one can find out that the execution times of all tasks increase with task periods. For instance, in the time interval $t = 3.5-4$ s, the execution times of two running tasks (i.e. the first two tasks) are clearly larger under EEAFS than under opDVS. According to the principle of DVS, larger execution times mean lower CPU speeds, which yield in turn lower energy consumption. It can be outlined that EEAFS is remarkably more efficient in reducing energy consumption than opDVS.

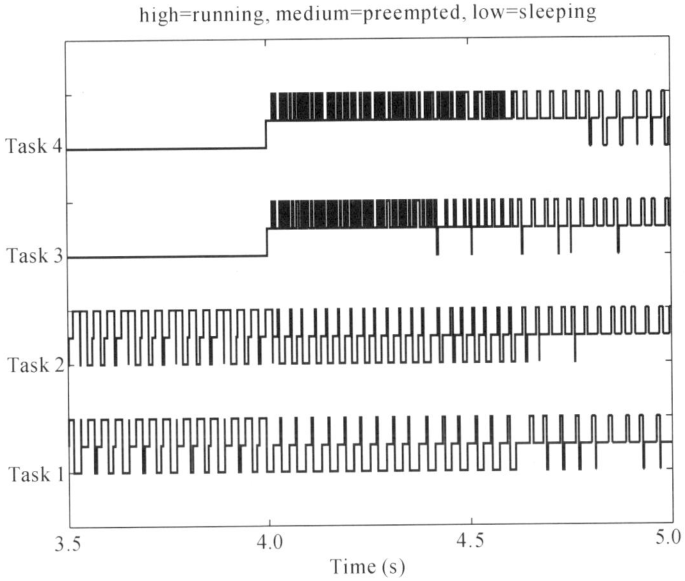

Fig. 6. 10 CPU schedule under EEAFS-1

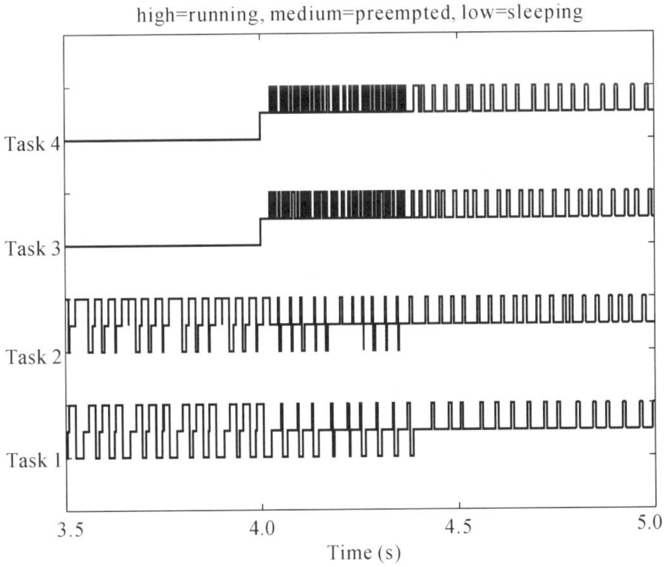

Fig. 6. 11 CPU schedule under EEAFS-2

QoC

Figs.6.12, 6.13 and 6.14 show the system responses of each control loop under different schemes, respectively. All four control loops achieve satisfactory performance under every scheme. For each loop, the QoC under different schemes is quite close.

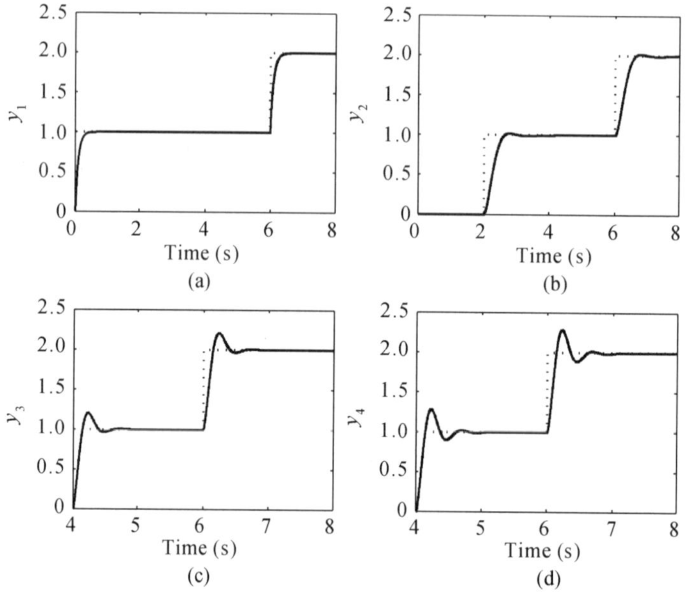

Fig. 6.12 System responses under opDVS
(a) Loop 1; (b) Loop 2; (c) Loop 3; (d) Loop 4

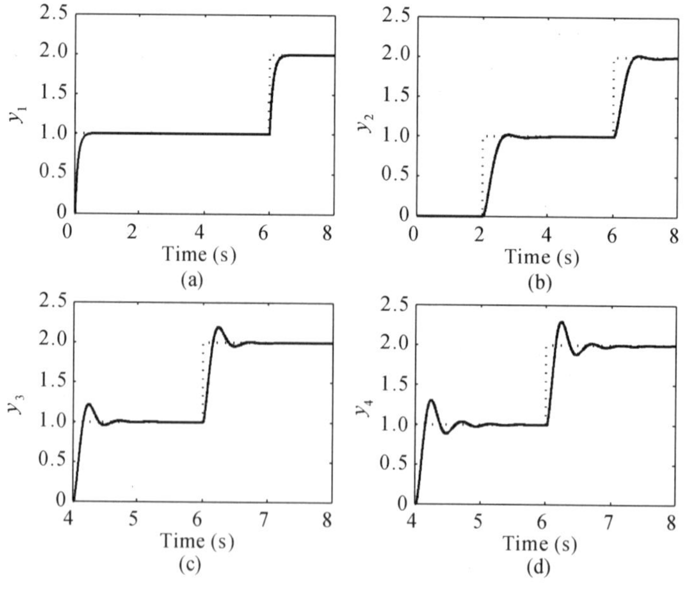

Fig. 6.13 System responses under EEAFS-1
(a) Loop 1; (b) Loop 2; (c) Loop 3; (d) Loop 4

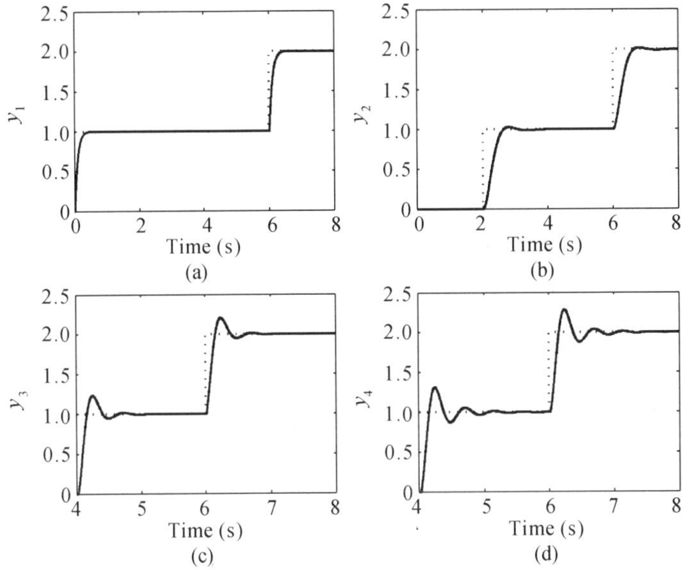

Fig. 6. 14 System responses under EEAFS-2
(a) Loop 1; (b) Loop 2; (c) Loop 3; (d) Loop 4

Fig. 6.15 depicts the total control cost J_{SUM} of the system. It is clear that the three schemes deliver comparable overall control performance. The accumulative total control costs are 1.309, 1.396 and 1.443, respectively. Compared with the opDVS case, the total control costs under EEAFS-1 and EEAFS-2 increase by 6.7% and 10.2%, respectively.

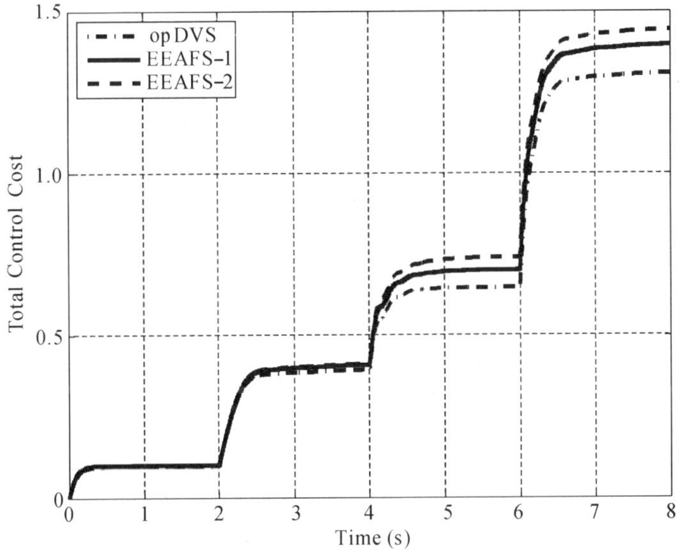

Fig. 6. 15 Total control costs under different schemes

To summarize this set of experiments, both the exponential and the linear EEAFS schemes can achieve significant additional energy consumption reduction over the optimal pure DVS scheme, while delivering comparable control performance. The linear EEAFS algorithm is more aggressive in energy saving than the exponential EEAFS algorithm (with β = 40), but with slightly degraded QoC. Overall, these two algorithms perform comparably. In the following experiments, unless otherwise specified, only the exponential EEAFS algorithm will be studied thanks to its higher flexibility in design.

6.5.2 Experiment Ⅱ: Different Design Parameters

Next the performance of EEAFS with different design parameters is assessed. In particular, the effect of different β values on EEAFS is studied. As Fig.6.4 indicates, the setting of β determines how to adjust the sampling periods, thus affecting the feedback scheduling performance of EEAFS.

The simulation pattern remains the same as in Experiment I. The set of β values chosen for simulations is {1, 10, 20, 40, 60, 80, ∞ }, where EEAFS with $\beta \to \infty$ implies a scheme similar to the dynamic solution in [LEE04].

Fig.6.16 shows the CPU energy consumption under EEAFS with different β values. The average CPU energy consumption E_{AVG} in different cases is summarized in Table 6.2. An obvious result is that the energy expenditure increases with β, which implies that smaller β values are more beneficial to energy saving. Even in the worst case where $\beta \to \infty$, however, EEAFS is still able to reduce 41.7% additional energy consumption on average over opDVS.

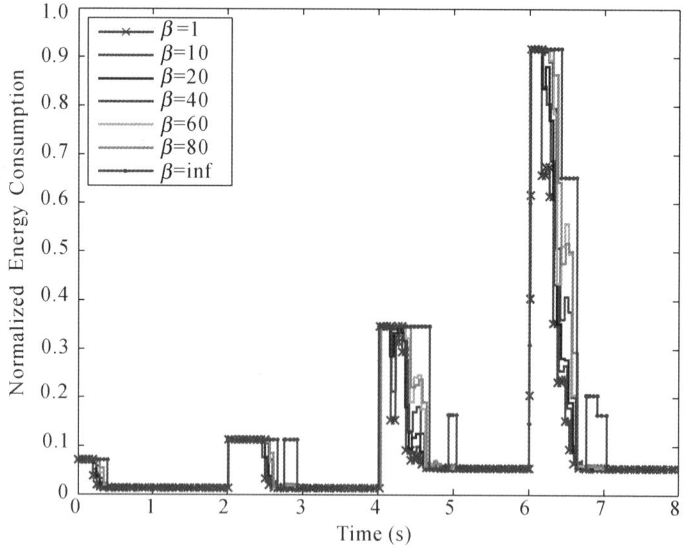

Fig. 6. 16 Normalized energy consumptions with different parameters

Table 6.2 Average energy consumptions and total control costs with different parameters

	$\beta=1$	$\beta=10$	$\beta=20$	$\beta=40$	$\beta=60$	$\beta=80$	$\beta\to\infty$
$E_{AVG}(\%)$	9.80	10.3	10.3	10.9	12.5	13.0	16.2
J_{SUM}	1.440	1.456	1.393	1.396	1.360	1.432	1.404

As Fig.6.17 shows, for each control loop, the difference between the accumulated control costs with different β values is minor. Also given in Table 6.2 are the total control costs of the system with different β values. The overall control performance under EEAFS with different β values is quite comparable.

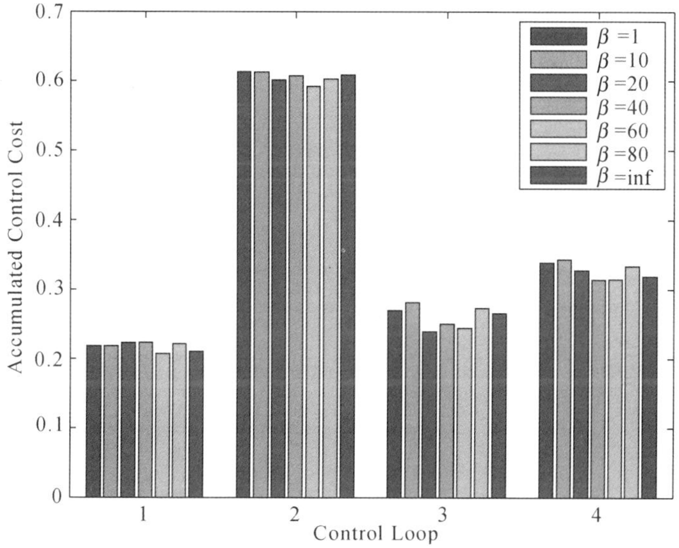

Fig. 6.17 Accumulated loop control costs with different parameters

In general, decreasing β reduces energy expenditure, but may possibly cause (minor) control performance degradation. Indeed, the choice of β corresponds to a trade-off between low energy consumption and high QoC. Fortunately, the EEAFS scheme performs pretty well for different β values in a very large range, e.g. from 1 to ∞, as shown in Table 6.2. Consequently, it is not difficult to choose an appropriate β during the design of EEAFS.

Besides β, design parameters such as e_{min} and e_{max} also affect the performance of EEAFS. Their effects will be studied by means of simulations and/or real experiments in our future work.

6.5.3 Experiment Ⅲ: Different Perturbation Intervals

As mentioned in Section 6.4, the performance of EEAFS in saving energy depends, to some degree, on the perturbations posed on each control loop, since the task periods are adapted to control errors that are highly related to the perturbations. Studied next is the performance of EEAFS when the control loops encounter perturbations with different frequencies.

In this set of simulations, all loops start running from $t = 0$, and are perturbed by step input changes at the same time. The Perturbation Interval (PI) is set to 1, 2, 4 and 6 s, respectively, for each run. The whole simulation lasts 12 s every time.

With $\beta = 40$, Figs.6.18 and 6.19 give the CPU energy consumptions and the total control costs of the system for different perturbation intervals.

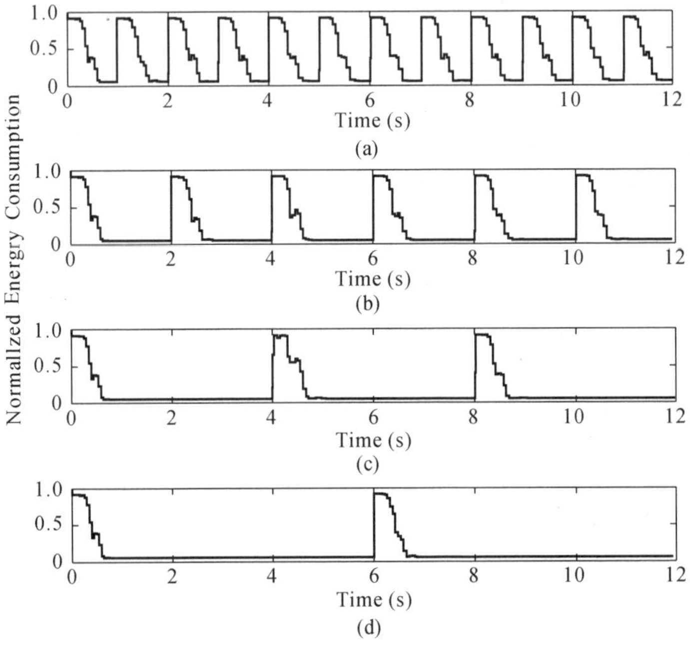

Fig. 6. 18 Normalized energy consumptions for different perturbation intervals (a) $PI = 1$ s; (b) $PI = 2$ s; (c) $PI = 4$ s; (d) $PI = 6$ s;

Table 6.3 summarizes the average energy consumptions and the total control costs for different perturbation intervals under both opDVS and EEAFS.

The CPU energy consumption under opDVS remains constant regardless of the changes in perturbation intervals. This is mainly due to that the system workload remains changeless during runtime, though the frequency of the perturbations varies. Not surprisingly, the total control costs increase with decreasing perturbation intervals, since the number of step input changes recorded in the simulations is in inverse proportion to the perturbation interval.

In contrast, the CPU energy consumption becomes highly related to how often the control loops encounter perturbations, when EEAFS is employed. It is apparent that larger perturbation intervals yield lower energy expenditure under EEAFS. When $PI = 1$ s, the average energy consumption under EEAFS is 48.0%, which is 43.8% less than under opDVS. As PI increases, the energy consumption under EEAFS decreases accordingly and as a consequence, the additional energy consumption reduction over opDVS increases. When PI is set to 6 s, the average energy con-

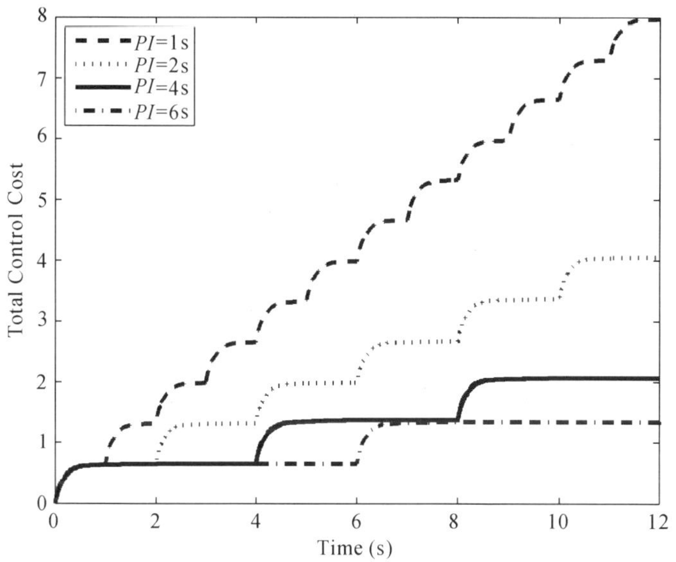

Fig. 6.19 Total control costs for different perturbation intervals

sumption under EEAFS reduces to 11.7%, which is 80.1% less than under opDVS. The advantages of EEAFS become more paramount as the perturbations posed on control loops become lighter, i.e. when the frequency of perturbations is smaller.

By observing the total control costs of the four control loops, it can be seen that the degradation of QoC under EEAFS over opDVS is always minor for different perturbation intervals. The relative increase of the total control cost remains within 7%. This argues that EEAFS and opDVS deliver comparable control performance.

Table 6.3 Average energy consumptions and total control costs with different perturbation intervals

PI (s)	E_{AVG}(%)			J_{SUM}		
	opDVS	EEAFS	Decrease	opDVS	EEAFS	Increase
1	91.8	48.0	43.8	7.591	7.978	5.1%
2	91.8	26.0	65.8	3.874	4.065	4.9%
4	91.8	15.9	75.9	1.952	2.075	6.3%
6	91.8	11.7	80.1	1.302	1.357	4.2%

6.5.4 Experiment IV: Discrete Voltage Levels

The above experiments are based on the assumption that the CPU speed could be scaled continuously and α_{min} be sufficiently small. In this set of simulations, the performance of EEAFS with $\beta = 40$ in the case of processors with discrete voltage levels is assessed. For simplicity of description, in the following the term *ideal* EEAFS is used to denote the EEAFS scheme for continuous variable voltage processors (also called *ideal* processors), while *discrete* EEAFS representing the EEAFS scheme that incorporates the methods presented in Section 6.4.4 to deal with limited allowable voltage levels.

The simulation pattern used here is the same as in Experiment III. The perturbation interval is set to 1 s. Consider four different processors that support the following normalized voltage levels:

(1) CPU-1: {0.5, 0.75, 1.0} [PIL01];

(2) CPU-2: {0.25, 0.5, 0.75, 1.0} [ZHU05];

(3) CPU-3: {0.36, 0.55, 0.64, 0.73, 0.82, 0.91, 1.0} [PIL01];

(4) CPU-4: {0.285, 0.333, 0.380, 0.428, 0.476, 0.523, 0.571, 0.619, 0.666, 0.714, 0.761, 0.809, 0.857, 0.904, 0.952, 1.0} [SOR05].

Fig.6.20 depicts the energy consumption of different processors. In contrast to Fig.6.18, when the processor supports only limited allowable voltage levels, the instantaneous samples of normalized energy consumption are also discretized. If CPU-1 is used, for example, the energy consumption can only take on values of {100%, 56.3%, 25%}. As Fig.6.21 shows, the difference in QoC of each loop caused by the use of different processors is quite minor.

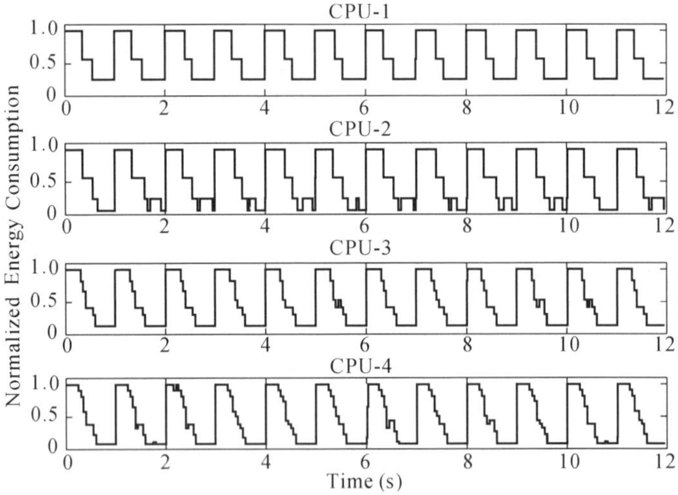

Fig. 6.20 Normalized energy consumptions of different processors

The results of this set of experiments are summarized in Table 6.4, where the simulation results for ideal EEAFS are also given for the purpose of comparison.

Table 6.4 Average energy consumptions and total control costs with different processors

	Ideal EEAFS	Discrete EEAFS			
		CPU-1	CPU-2	CPU-3	CPU-4
E_{AVG}(%)	48.0	59.5	54.0	50.2	46.7
J_{SUM}	7.978	7.730	8.005	7.833	8.043

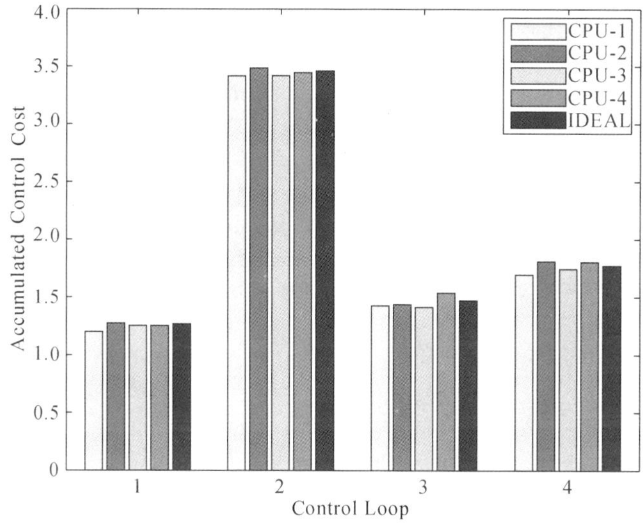

Fig. 6.21 Accumulated loop control costs with different processors

Compared to the ideal case in which the CPU speed can be scaled continuously, the limitation of allowable voltage/speed levels will possibly cause the CPU energy consumption to increase. The main reason behind this is that the CPU speed assigned to discrete variable voltage processors may be lower than the optimal/ideal value in some circumstances. For processors with only a small number of allowable voltage levels, in comparison with the ideal EEAFS, the discrete EEAFS scheme performs worse in energy saving, whereas the QoC may be improved.

When the first processor with only three voltage levels is used, for instance, the average energy consumption increases by 11.5%, and meanwhile, the total control cost decreases by 3.1%. The increase of energy consumption results from the significant difference between the actually assigned CPU speeds and the corresponding ideal values, which occurs as a result of the voltage level quantization. On the other hand, the improvement of QoC mainly benefits from the resource reclaiming mechanism.

The minimum allowable voltage level of the processor also affects the performance of EEAFS. For example, the minimum possible speed of CPU-1 is 0.5, which poses a natural constraint on the energy expenditure, i.e. the CPU energy consumption can by no means be less than 25%, even when the system workload is lower than 0.5.

It can be found from Table 6.4 that the performance of the discrete EEAFS in energy saving approaches that of the ideal EEAFS as the number of allowable voltage levels increases. The larger the number of allowable voltage levels, the better the performance of EEAFS in energy saving. When CPU-1 that supports 3 voltage levels is employed, the average energy consumption under EEAFS increases by 11.5% in comparison with the ideal case. However, their difference deceases to 2.2%, which indicates comparable performance in energy saving, when CPU-3 supporting 7 voltage levels is used. With regard to QoC, EEAFS performs comparably for different processors including the ideal one.

It is interesting that the average CPU energy consumption becomes 1.3% less than that of the ideal case when CPU-4 is used. The major reasons behind this result are twofold. First, according to the rational of EEAFS, the CPU energy expenditure is heavily related to the variations in the control error of each loop. In the two cases where the ideal processor and CPU-4 are used respectively, the QoC of each loop is different, which may possibly lead to different energy consumption. Second, CPU-4 supports a sufficiently large number of (i.e. 16) allowable voltage levels, which makes the difference between the ideal processor and CPU-4 insignificant. Therefore, it is possible for CPU-4 to consume even less energy than the ideal processor.

It can be outlined from the above results that, when applied to processors with limited voltage levels, EEAFS can still perform well in energy saving, while delivering good QoC. As a general rule, the larger the number of allowable voltage levels, the lower the energy consumption. When the number of allowable voltage levels is relatively small, more computing resources will be expended by control loops for the purpose of improving QoC. However, the control performance improvement is at the cost of higher energy expenditure. Since the basic idea of EEAFS is to reduce energy consumption without jeopardizing control performance, processors with more allowable voltage levels are preferable from the viewpoint of improving the energy efficiency of the system.

6.6 Summary

The objective of this chapter is to achieve further energy saving over the optimal pure DVS scheme. To meet this purpose, the behaviour of control systems with variable sampling periods has been studied, and an Enhanced Energy-Aware Feedback Scheduling scheme exploiting the methodology of graceful degradation has been presented. While adjusting task execution times through manipulating the CPU speed based on the DVS technique, this scheme adapts each task's period to relevant control performance, which enables flexible timing constraints. Since the task periods are enlarged provided that the QoC is not jeopardized, higher energy efficiency is achieved. Extensive simulation experiments have demonstrated the merits of the proposed approach under different circumstances.

The major contributions of this chapter can be briefly outlined as follows:

(1) From the perspective of feedback scheduling, this chapter presents a novel feedback scheduling scheme that features simultaneous adjustment of task execution times and periods.

(2) For the area of DVS, this chapter proposes to exploit application adaptation to enhance the performance of (the optimal) pure DVS schemes.

(3) For the area of real-time control, this chapter presents a highly cost-effective method for improving the energy efficiency of embedded controllers.

References

[ARZ99] K.E. Årzén. A Simple Event Based PID Controller. *Proc. 14th IFAC World Congress*, Beijing, China, pp.423 – 428, 1999.

[AYD04] H. Aydin, R. Melhem, D. Mosse, P. Mejia-Alvarez. Power-Aware Scheduling for Periodic Real-Time Tasks. *IEEE Trans. Computers*, Vol. 53, No. 5, pp.584 – 600, 2004.

[GRA04] I.A. Gravagne, J.M. Davis, J.J. Dacunha, R.J. Marks. Bandwidth Reduction for Controller Area Networks Using Adaptive Sampling. *IEEE Int. Conf. on Robotics and Automation*, Vol. 5, pp.5250 – 5255, 2004.

[LEE04] H.S. Lee, B.K. Kim. Design of Digital Control Systems with Dynamic Voltage Scaling. *10th IEEE Real-Time and Embedded Technology and Applications Symposium (RTAS'04)*, 2004.

[LIN03] Q. Lin, P.C.Y. Chen, P.A. Neow. Dynamical Scheduling of Digital Control Systems. *IEEE Int. Conf. on Systems, Man and Cybernetics*, Vol. 5, pp.4098 – 4103, 2003.

[MAR02] P. Marti, G. Fohler, K. Ramamritham, J.M. Fuertes. Improving Quality-of-Control Using Flexible Time Constraints: Metric and Scheduling Issues. *Proc. 23rd IEEE RTSS*, Austin, TX, USA, pp.91 – 100, 2002.

[MES06] B. Messner, D. Tilbury. Control Tutorials for MATLAB and Simulink, http://www.library.cmu.edu/ctms/, 2006.

[PER98] T. Pering, T. Burd, R. Brodersen. The Simulation and Evaluation of Dynamic Voltage Scaling Algorithms. *Proc. of the ACM Int Symposium on Low Power Electronics and Design*, New York, NY, USA, pp. 76 – 81, 1998.

[PIL01] P. Pillai, K.G. Shin. Real-Time Dynamic Voltage Scaling for Low Power Embedded Operating Systems. *Proc. 18th ACM Symposium on Operating Systems Principles*, Banff, Alberta, Canada, pp.89 – 102, 2001.

[SIN01] A. Sinha, A.P. Chandrakasan. Energy Efficient Real-Time Scheduling. *Proc. Int. Conf. Computer Aided Design*, pp.458 – 463, 2001.

[SOR05] A. Soria-Lopez, P. Mejia-Alvarez, J. Cornejo. Feedback Scheduling of Power-Aware Soft Real-Time Tasks. *Proc. 6th Mexican Int. Conf. on Computer Science (ENC'05)*, pp.266 – 273, 2005.

[XIA06] F. Xia, Y.X. Sun. An Enhanced Dynamic Voltage Scaling Scheme for Energy-Efficient Embedded Real-Time Control Systems. *Lecture Notes in Computer Science*, Vol. 3983, pp.539 – 548, 2006.

[XIA07a] F. Xia, Y.C. Tian, Y.X. Sun, J.X. Dong. QoC-Aware Power Management in Embedded Control Systems with Multiple-Voltage Processors. Submitted, 2007.

[XIA07b] F. Xia, Y.C. Tian, W.H. Zhao, M.O. Tadé, Y.X. Sun. Enhanced Energy-Aware Feedback Scheduling of Embedded Control Systems. Submitted, 2007.

[ZHA06] W.H. Zhao, F. Xia. An Efficient Approach to Energy Saving in Microcontrollers. *Lecture Notes in Computer Science*, Vol. 4186, pp. 595 - 601, 2006.

[ZHU05] Y. Zhu. Dynamic Voltage Scaling with Feedback EDF Scheduling for Real-Time Embedded Systems. Ph.D Thesis, North Carolina State University, 2005.

PART IV

BANDWIDTH ALLOCATION

Chapter 7

Integrated Feedback Scheduling

With traditional open-loop network scheduling methods, the performance of networked control systems may be significantly impacted by limited network bandwidth and time-varying workloads. This chapter considers multi-loop networked control systems that operate in dynamic and uncertain environments. Exploiting the methodology of integrating control and scheduling, we attempt to maximize the overall control performance through allocating online available network bandwidth resources among control loops.

An integrated feedback scheduling scheme is presented on the basis of codesign of control and network scheduling. It adapts both the sampling periods and the priorities of control loops at runtime, in response to changes in available network resource. A cascaded feedback scheduling algorithm is developed to alter the sampling periods so that the deadline miss ratio is controlled at a desired level and available bandwidth is distributed in an optimal fashion. For priority modification, a direct feedback scheduling algorithm is used to re-assign control loops' priorities according to their actual control performance. This will improve the overall control performance. It is demonstrated by simulation experiments that integrated feedback scheduling can maximize resource utilization in the case of light workload, and achieve graceful performance degradation under overload conditions. In comparison with the traditional design method, it can significantly improve the overall QoC of the system.

7.1 Introduction

Throughout the last decade networked control systems (NCSs) [LI06, TIP03, XIA04a, XIA04b, XIA04c, XIA05b, ZHA06a, ZHA06b] are undoubtedly one of the hottest research topics in the field of control. An NCS usually uses a control network to connect node devices such as sensors, controllers and actuators that are geographically distributed. These nodes exchange information over the shared bus, thus functioning control of physical processes.

One could refer to Fig.2.1 for the typical architecture of an NCS. Compared with point-to-point control systems, networked control systems are advantageous in terms of simple and fast implementation, ease of system maintenance, and increased system flexibility and dependability, thanks to substantial reduction in wiring. Nowadays various NCSs closed over control networks have been widely deployed in many areas such as process control, factory automation, and automotive electronics.

Regardless, the use of shared communication networks makes it hard to design and analyze control systems. The resulting performance of the system depends heavily on temporal attributes such as network-induced delay, packet loss, and jitter. To attack these problems that exist in NCSs, researchers have suggested many solutions, including, e. g. methods based on either controller design or network design. Although significant progress has been made in networked control theory and practice over the years [LI05, TIP03], NCSs increasingly feature today dynamic constraints on the available network bandwidth, which raises new challenges in NCS design.

7.1.1 Motivation

From the communication perspective, to satisfy the requirements of control applications, it is often mandatory for control networks to provide deterministic real-time communication. This poses a technical limitation on the maximum possible transmission rate that control networks can offer. For example, the CAN bus has a maximum transmission rate of 1 Mbps [BOS91]. Control networks with much higher data rates are available now, but the accompanying cost could be very high. Therefore, it is common in real-world applications that the bandwidth of control networks is limited [BEN06]. Furthermore, there is no foreseeable change of this situation in the next few years due to both technical and economical reasons. On the other hand, to meet diverse system requirements concerning, e.g. reconfigurability, extensibility and flexibility, many NCSs have to operate under changing environments. As a result, the network workload may vary over time.

As mentioned in Chapter 1, a natural result of limited network bandwidth and variable workload is the uncertainty in available communication resource. In terms of temporal properties, it yields unpredictable communication delays, packet loss, and jitter, which may possibly deteriorate the QoC, and even jeopardize system stability in extreme circumstances. For a certain control loop, the availability of time on a shared network for communications between the sensor, the controller, and the actuator is an important factor that determines the resulting control performance. Therefore, the overall performance of an NCS that consists of multiple control loops depends not only on the design of control algorithms, but also on the allocation and scheduling of the shared network bandwidth [BEN06, XIA05a].

The research approach of most existing work in the area of NCS could roughly be outlined as follows. First, the temporal behaviour of the control network is analyzed, either theoretically or experimentally; and then the effect of the use of the communication network will be dealt with via either controller design or network design. Most often these methods are overly idealized, e.g. in modelling timing attributes such as communication delay, packet loss, jitter, etc. This makes them not applicable to dynamic environments featuring uncertainty in resource availability, or too complex to implement in real applications. In order to manage the performance bottleneck within NCSs more efficiently, in recent years some researchers have paid special attention to the problem of network scheduling in the context of networked control, e.g. [WAL01, YEP03]. However, these papers show little concern with the codesign of feedback control and real-time scheduling.

Zhang [ZHA01] and Branicky, *et al.* [BRA02] introduce the idea of integrating control and scheduling into networked control systems. They also give some fundamental methods. Following this codesign methodology, [ALH05, HE04, VEL04] propose different algorithms for adapting sampling periods. A common feature of these methods is that they are dependent on pre-set upper bounds for network utilization. However, in network scheduling it is very hard, if not impossible, to determine the exact upper bound for network utilization that guarantees system schedulability. Most often pessimistic utilization setpoints are used in these methods in order not to jeopardize the system schedulability. Consequently, an intrinsic disadvantage with these methods is that they cannot make full use of the available network resource. Additionally, none of them deals with online modifi-cation of node priorities.

7.1.2 Contributions

This chapter is devoted to maximizing the overall control performance of NCSs by means of flexible management of network resources, which is in contrast to traditional solutions in this area. Considering multi-loop NCSs subject to bandwidth limitations and workload fluctuations, we present an Integrated Feedback Scheduling (IFS) scheme [XIA06b, XIA07b] that exploits the emerging methodology of codesign of feedback control and network scheduling. Assuming priority-driven communi-cation protocols, we develop the mapping between network scheduling and CPU scheduling and describe the codesign problem within NCSs. To allocate available bandwidth resources flexibly, we dynamically adapt both sampling periods and priorities of control loops simultaneously. The overall control performance is expected to be optimized by means of making optimal use of network resource.

For the adjustment of sampling periods, we present a cascaded feedback scheduling algorithm [XIA07a] that consists of the following two procedures. First, the total network utilization of all control loops is regulated using feedback control techniques, with the goal of keeping the deadline miss ratio at a desired (low) level. Then, the sampling periods are optimized subject to constraints on the total utilization obtained in the first procedure. On the one hand, this method is able to make full use of the network resource even when the system workload is light. On the other hand, by employing this method the system can achieve graceful QoC degradation under overload conditions. To make better use of the communication resource, we propose to dynamically modify the nodes' priorities according to the control loops' actual performance, along with the adjustment of sampling periods. In general, a control loop with worse performance will be assigned a higher priority. In this way the communication delay within this loop will be reduced and the possibility of the corresponding packets missing deadline decreases. Intuitively this benefits further improvement of the overall control performance of the system. In contrast to traditional open-loop scheduling methods for NCSs, our approach features closed loop scheduling of network bandwidth. We also conduct extensive simulation experiments to highlight the effectiveness and merits of the proposed approach.

The rest of this chapter is structured as follows. Section 7.2 describes the model of the system to be considered, and describes the problem of network scheduling in the context of NCS from a viewpoint of integrated control and scheduling. In Section 7.3 we present the integrated feedback scheduling scheme for NCSs operating in dynamic environments. The architecture of the scheme is illustrated. Then the algorithms used for sampling period adjustment and priority modification are given, followed by discussions on some critical design issues. Section 7.4 evaluates the performance of the proposed approach via simulation experiments under different scenarios. Finally, we summarize this chapter in Section 7.5.

7.2 Problem Statement

7.2.1 System Model

Consider a networked control system shown in Fig.7.1, where N independent control loops share one control network. In this network a priority-based MAC protocol is employed. Commonplace examples of this type are CAN and DeviceNet. According to the principle of this type of communication protocols, every node with packets to send, i. e. so-called communication entity will be assigned a specific priority level. In general circumstances, the priority of each node is unique, which means that packets from different nodes are associated with different priority levels. A node with a packet to send waits until the network is idle and then commences to transmit. In the case of network access collisions, the system will decide which packet will be transmitted according to their priority levels. As usual the packet with the highest priority will be transmitted successfully. It should be noted that the proposed approach is also applicable to networked control systems with interfering traffic, though Fig.7.1 does not include any disturbing nodes for the sake of simple description.

In this system each control loop is composed of a sensor, a controller, and an actuator, in addition to a controlled process. The controller and the actuator are connected directly, which implies that only the sensor in the control loop needs to use the control network to deliver sample data to the corresponding controller. With

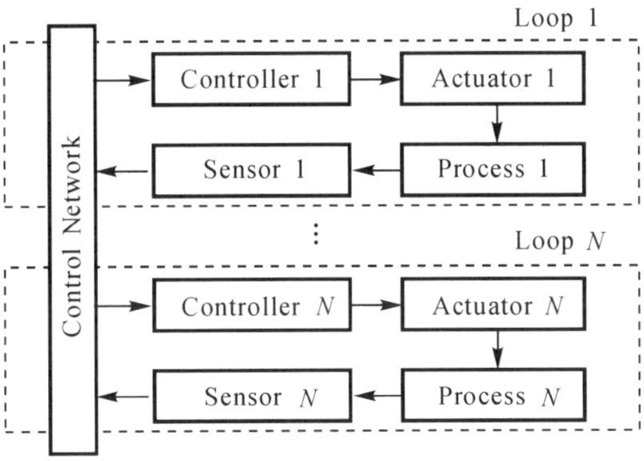

Fig. 7.1 A typical multi-loop networked control system

this system architecture, we view a sensors node's priority as the priority of the relevant control loop. All controllers and actuators are event triggered, while sensors are time-triggered. In a single control loop, the sensor collects a sample of the output of physical process at the beginning of every sampling period, and then sends it to the controller via the control network after getting the right to access the network. Upon receiving the sample data, the controller starts to execute the control algorithm immediately, produces control input, and then outputs it into the actuator. Finally the actuator acts on the controlled physical process according to the control input. Assume that in this process the processing delay of the sensor and actuator and the execution time of the control algorithm are relatively so small that they could be neglected. In this context the control delay is approximately equal to the communication delay which encompasses both the waiting delay and the transmission delay.

No method is used in the control loops to compensate for delay, packet loss, and jitter. To improve the feedback scheduling performance, however, controller parameters will be updated accordingly when sampling periods are changed.

Without loss of generality, we assume in the above system:

(1) Data communications over the control network are reliable in that no errors (e.g. bit error) exist in the process of transmission;

(2) Sample data is delivered in the form of single packets, which means that every sample will be treated as one data packet while being transmitted over the network;

(3) The sensor and the actuator in one loop hold a precisely synchronized clock. Since only the sensors are time-triggered in the system, this assumption is not a necessity. The purpose of making it here is simply for exact calculation of the deadline miss ratio.

7.2.2 Control Network Scheduling

In the NCS shown in Fig.7.1 the control network is shared by multiple communication entities. Therefore, it is necessary and important to allocate the network bandwidth properly, particularly when the transmission rate is limited. In the following we try to formulate the problem of control network scheduling from the perspective of real-time CPU scheduling, and make a mapping between them.

Generally speaking, the scheduling of networks is similar to the problem of real-time CPU scheduling [ZHA01]. Both of them consider how to distribute shared resources among a set of concurrent tasks, which are often subject to real-time constraints, either hard or soft. Obviously, the shared resource to be distributed is no longer the CPU time, but the network bandwidth in NCSs. The tasks to execute in this context will not be software programs as in CPU scheduling, but messages (i.e. data packets) over the network. Accordingly, the meaning of task execution changes from running programs to transmitting data packets. Based on these observations, it could be found that existing real-time scheduling theory and methods may be applied to control network scheduling by redefining relevant task attributes. In this chapter we define the following timing attributes for sample data messages in the context of NCS:

(1) Period h_i: the period for generating a new sample data packet in the i-th control loop, which is equal to the relevant sampling period;

(2) Relative deadline d_i: equal to h_i;

(3) Execution time c_i: the transmission delay of a data packet, excluding the waiting delay;

(4) Priority p_i: the sensor's priority in the ith loop;

(5) Network utilization U_i : *equal to* c_i/h_i.

Table 7.1 compares some important attributes of CPU scheduling in embedded control systems and network scheduling in networked control systems. Traditional dynamic-priority scheduling algorithms (e.g. EDF) are theoretically applicable to control network scheduling based on this mapping. Due to practical difficulty in implementation, however, fixed-priority scheduling methods are used in almost all real-world NCSs.

Table 7.1 Attributes of CPU scheduling and network scheduling

	CPU scheduling	Network scheduling
Task	Control program	Data packet
Period (relative deadline)	Sampling period of control program	Sampling period of sensor
Execution time	Computation time of control algorithm	Transmission time of data packet
Priority	Program's priority	Sensor's priority
Utilization	CPU utilization	Network utilization
Preemptivity	Preemptive	Non-preemptive

As shown in Table 7.1, the most notable difference between CPU scheduling and network scheduling is that in general the execution of programs on processors is preemptive, whereas the transmission of data over networks is not. Once a data transmission starts, it will continue until done, and will never be suspended because of new transmission requests with higher priorities. For this reason, Theorem 2.1 is not applicable to the case of network scheduling. To analyze the schedulability of NCSs, real-time scheduling theory and methods for non-preemptive tasks are required. In fact, it is quite easy to recognize the following sufficient condition for NCS schedulability analysis in accordance with the theory by Sha, *et al.* [SHA90].

Theorem 7.1 For a networked control system composed of N independent control loops, where an ideal priority-based control network is used, the system is schedulable with RM algorithm if Eq.(7.1) is satisfied.

$$\frac{c_1}{h_1} + \frac{c_2}{h_2} + \cdots + \frac{c_i}{h_i} + \frac{b_i}{h_i} \leqslant i(2^{\frac{1}{i}} - 1), \quad \forall i = 1, \cdots, N, \tag{7.1}$$

where $h_1 \leqslant h_2 \leqslant \cdots \leqslant h_N$, b_i is the worst case blocking time of task i, i.e. $b_i = \max_{n=i+1,\cdots,N} c_n$.

Based on the above schedulability constraint, the network scheduling problem in NCSs can be stated informally within the framework of feedback scheduling as follows: *In the presence of limited bandwidth and variable workload, dynamically allocate available communication resources among multiple control loops so that the overall control performance is maximized*. In the next section we will present an integrated feedback scheduling approach to address this problem.

7.3 Integrated Feedback Scheduling

In contrast to traditional open-loop network scheduling methods, this chapter aims to develop a closed loop dynamic network bandwidth allocation method. Based on the above description, we propose to adapt simultaneously the sampling periods and the priorities of concurrent control loops by using different feedback scheduling algorithms at different levels. In this way, flexible QoC management will be facilitated by means of codesign of feedback control and network scheduling. In this section we first describe the architecture of integrated feedback scheduling, illustrating the main idea behind this method; then we will present, respectively, the feedback scheduling algorithms for period adjustment and priority modification; and finally, some critical design issues will be discussed.

7.3.1 Architecture

To enable feedback scheduling, we must first make proper choices of relevant variables. According to the principle of feedback control, the *controlled variable* and the *manipulated variable* are the two most important types of variables that need to be defined in the feedback scheduling system. In terms of real-time scheduling, there are generally two options for the controlled variable, network utilization and deadline miss ratio (or miss ratio in short).

For feedback scheduling of embedded processors, in most cases both utilization (see Chapters 3 to 6) and deadline miss ratio can be chosen as the controlled variable. In network scheduling, however, it is very hard, if not impossible, to determine the exact schedulable utilization upper bound, whereas in preemptive CPU scheduling this is pretty easy. It consequently becomes difficult to choose an appropriate network utilization setpoint, especially when the network workload varies with time.

Actually, the deadline miss ratio is a good choice for the controlled variable when dealing with network scheduling. Feedback scheduling based on deadline miss ratio does not depend on knowledge about the schedulable network utilization upper bound, which avoids the difficulty with network schedulability analysis. Moreover, it is intuitive that controlling deadline miss ratio at a specific (relatively low) level will certainly keep the actual network utilization close to the highest possible level, regardless of changes in system workload. Because there is always a certain stability margin in practical control systems design, real-world control systems can tolerate packet losses (and of course deadline misses) to a certain degree. Therefore, in this chapter we select deadline miss ratio as the output of the feedback scheduling system. As regarding the manipulated variable, since the transmission time of sample data packets cannot be intentionally adjusted, the sampling periods of the control loops become a natural choice.

Fig.7.2 depicts the architecture of integrated feedback scheduling. An outer loop that implements feedback scheduling is introduced in addition to the original control loops shown in Fig.7.1. The feedback scheduler consists of two main components: *period adjustment* and *priority modification*. Just as their names imply, the period adjustment module is responsible for dynamically adjusting the sampling periods of control loops according to workload level and system performance, and the priority modification module re-assigns the sensors' priorities according to actual performance of control loops. Clearly, the inputs to the integrated feedback scheduler include the actual deadline miss ratio and its setpoint, and the measured QoC in terms of some performance metric, while the outputs are the sampling period and the priority of each control loop.

The feedback scheduler developed in this chapter is time-triggered. At every invocation instant, the scheduler gathers the current deadline miss ratio and the QoC of control loops. Based on the setpoint for deadline miss ratio, new sampling periods will be produced. The priorities of control loops will also be modified when necessary.

Intuitively, the actual network utilization could be maintained around the highest possible level through adapting sampling periods, as long as the deadline miss ratio setpoint is not zero. Due to this non-zero deadline miss ratio setpoint, however, deadline misses inevitably exist when the system is in steady states, no matter at what level the workload is. Naturally we can keep the deadline miss ratio low through specifying a relatively small setpoint for deadline miss ratio. Unfortunately, in NCSs that use fixed-priority scheduling methods, most of the deadline misses will happen in control loops whose priorities are relatively low. This may possibly result in remarkable impact on these control loops' performance. This problem will be addressed in the priority modification module.

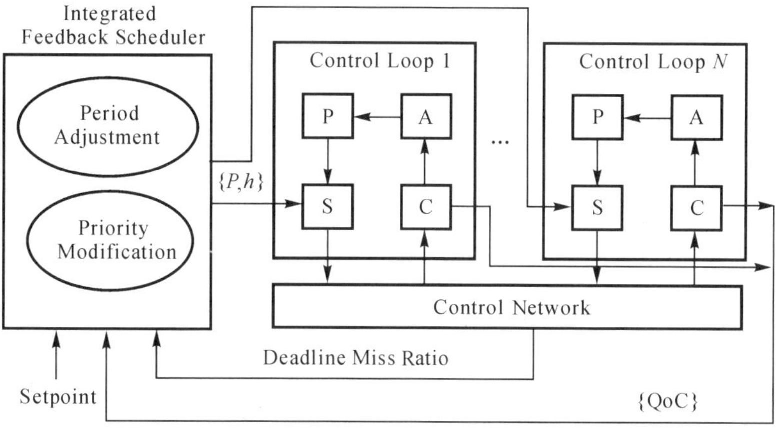

S:Sensor, C:Controller, A:Actuator, P: Process

Fig. 7.2 Architecture of integrated feedback scheduling

7.3.2 Period Adjustment

The major role of the period adjustment module is to adapt the control loops' sampling periods to current system state in a way that the actual deadline miss ratio is maintained at a desired level and the available resource is reasonably distributed. A natural method to construct this module based on feedback control theory is using a specific control algorithm to obtain sampling periods directly based on the difference between actual deadline miss ratio and its setpoint. In [XIA06b] we explore such a method. In this chapter we present a new cascaded feedback scheduling method to determine sampling periods. In contrast to our method in [XIA06b], the approach proposed here does not need to rescale all sampling periods in proportion. It is therefore more generalized. Since this approach allocates communication resources in accordance with actual control performance, it may yield better overall system performance.

We propose to produce the sampling periods with two consecutive steps, thereby forming a cascaded feedback scheduling algorithm. Firstly, use the classical PID control technique to calculate the total network utilization of all control loops in response to current control error of deadline miss ratio. Secondly, under this constraint of allowable network utilization, obtain new sampling periods while taking into account the control loops' actual performance.

Let T_{FS} be the invocation interval of the feedback scheduler. Deadline miss ratio $\rho(j)$ is defined as the ratio of the number of sample data packets that miss their deadlines to the total number of sample data packets generated in control loops in the time interval $[(j-1)T_{FS}, jT_{FS}]$, where j denotes the invocation instant of the feedback scheduler. Let ρ_r be the desired deadline miss ratio. At the jth instant, the total network utilization of control loops $U(0 \leqslant U \leqslant 1)$ is calculated as:

$$\Delta U(j) = K_P \cdot [ERR(j) - ERR(j-1)] + K_I \cdot ERR(j)\ , \tag{7.2}$$
$$U(j) = U(j-1) + \Delta U(j)$$

where K_P and K_I are essential coefficients for the PI control algorithm, ERR is the control error of deadline miss ratio.

The calculation of ERR is worth discussing. In most situations one could simply compute the control error as $ERR(j) = \rho_r - \rho(j)$ according to its general definition. However, deadline miss ratio is subject to saturation, that is, it could never be a negative. When the network workload is light, the deadline miss ratio will be zero all the time, regardless of changes in workload. On the other hand, a relatively small setpoint for deadline miss ratio is always preferable in order to minimize the impact of deadline misses on QoC of control loops. Consequently, in the case of light workload when almost no deadline is missed, the absolute value of control error will remain considerably small, no matter how low the level of workload actually is. This may cause the transient process of the feedback scheduling system to be very slow, thus impairing its dynamic behaviour.

To attack this problem, we here employ a deadzone-based control technique, and redefine the calculation of control error accordingly. From the control perspective, the deadline miss ratio should be as small as possible in favour of performance improvement. From the perspective of resource utilization, a non-zero deadline miss ratio implies that nearly full use of available resources has been made. Based on these observations, the system is considered to be steady if the deadline miss ratio falls into the interval $(0, \rho_r]$. Accordingly, if $0 < \rho(j) \leqslant \rho_r$, then the control error of deadline miss ratio is zero. To improve the dynamic behaviour of the feedback scheduling system, we use Eq.(7.3) to calculate $ERR(j)$.

$$ERR(j) = \begin{cases} \rho_r & \text{if } \rho(j) = 0 \\ 0 & \text{if } \rho_r \geqslant \rho(j) > 0\ . \\ -\rho(j) & \text{if } \rho(j) \geqslant \rho_r \end{cases} \tag{7.3}$$

It can be seen that the above equation enhances the performance of the feedback scheduler, which guarantees that the deadline miss ratio could come back into the steady region quickly, regardless of whether the system is underloaded without deadline miss or overloaded with severe deadline miss.

Once the total utilization $U(j)$ is obtained, we need to distribute it among control loops in a way that the overall control performance is optimized. Let J_i be the performance index for the ith control loop. The problem of finding the optimal sampling periods can be described as follows:

$$\begin{aligned} &\min_{h_1, \cdots, h_N} J = \sum_{i=1}^{N} w_i J_i \\ &\text{s. t.} \sum_{i=1}^{N} \frac{c_i}{h_i} \leqslant U(j) \end{aligned}\ , \tag{7.4}$$

where w_i is the weighting coefficient of loop i. It is apparent that the above equation gives a typical optimal feedback scheduling problem, the basic idea of which is to minimize the total control cost of the system through adjusting the sampling periods of control loops under the constraint of current total utilization. For details on optimal feedback scheduling, we refer the reader to Chapter 3. Recall that the bigger the control cost the worse the control performance.

Using the above formulation, the results of optimal sampling periods will tightly rely on the forms of the control loops' performance indexes. Generally speaking, J_i could be either stationary (i.e. independent of time) or dynamic (i.e. time-varying). In the time domain, there are commonly three kinds of options for J_i, i.e. infinite-time, finite-time and instantaneous ones. For instance, [EKE00, SET96] and Chapter 3 of this book employ infinite-time performance indexes; [CAS06] and [HEN05] exploit finite-time performance indexes; [MAR04, XIA05a, XIA06a] use instantaneous control error as performance index.

For the sake of simplicity, in this chapter we use the absolute instantaneous control error as the dynamic performance index for QoC, i.e.

$$J_i(j) = |e_i(j)|, \tag{7.5}$$

where e_i is the control error within the ith control loop. It should be noted that control errors of control loops are different from those of the feedback scheduling system.

With Eq.(7.5), Eq.(7.6) is used to compute the sampling period of loop i.

$$h_i = \frac{c_i}{U_i} = \frac{c_i}{U_{i,\min} + U'_i} = \frac{c_i}{U_{i,\min} + \left(U - \sum U_{n,\min}\right)\dfrac{w_i J_i}{\sum w_n J_n}} = \frac{c_i \sum_{n=1}^{N} w_n J_n}{\dfrac{c_i}{h_{i,\max}} \sum_{n=1}^{N} w_n J_n + w_i J_i \left(U - \sum_{n=1}^{N} \dfrac{c_n}{h_{n,\max}} \right)}, \tag{7.6}$$

where $h_{i,\max}$ is the maximum allowable sampling period, and $U_{i,\min} = c_i/h_{i,\max}$ is the corresponding minimum allowable utilization. Notice that Eq. (7.6) omits the indicator of feedback scheduler invocation instant (i.e. j) for the sake of notational simplicity.

Once the minimum utilization of each loop has been preserved, Eq.(7.6) will distribute the remaining fraction of bandwidth resource, i.e. $U(j) - \sum_{n=1}^{N} U_{n,\min}$, among control loops according to the values of $w_i J_i$. Control loops with worse performance will be assigned larger fractions of free network bandwidth, while the sampling periods of control loops with better performance will be more close to their maximum allowable values. In the extreme, if $J_i = 0$ indicating that the control loop is in a steady state, then it can be deduced from Eq.(7.6) that $h_i = h_{i,\max}$, that is, the sampling period of this loop is set to the maximum. With this bandwidth allocation scheme, we distribute the largest fraction of free communication resource to the control loop that needs it most. This benefits optimizing the overall control performance.

In Eq.(7.6) we assume that $\sum w_n J_n \neq 0$. In the case where all control loops are in steady states, i.e. $\sum w_n J_n = 0$, the free bandwidth will be distributed among control loops evenly. Furthermore, to avoid too big differences between the control loops' network utilization when $\sum w_n J_n$ is relatively small, which may be caused by the fact that some control loops are steady while the others are not, we introduce the following rule into the calculation of h_i:

$$h_i = \frac{c_i}{U_i} = \frac{c_i}{U_{i,\min} + U'_i} = \frac{c_i}{U_{i,\min} + \dfrac{U - \sum U_{n,\min}}{N}} \quad \text{if } \sum w_n J_n < \varepsilon, \tag{7.7}$$

where ε is a user-defined parameter.

7.3.3 Priority Modification

The reasons why we propose to modify the priorities of control loops online together with sampling period adjustment can be explained roughly in the following two aspects:

(1) To alleviate the effect of deadline misses. As mentioned previously, the feedback scheduling method based on deadline miss ratio control proposed in this chapter will unavoidably induce deadline misses in control loops, and the performance of control loops with low priorities will be influenced. To address this problem, the sensors' priorities are modified dynamically according to feedback information about actual control performance and assign lower priorities to control loops with better performance. In this way, the deadline misses are most likely to occur in control loops whose performance is the best, which is of benefit to reducing the impact of deadline misses on the overall QoC of the system.

(2) To alleviate the effect of delays. According to the principle of priority-based communication protocols, high-priority data packets will be transmitted earlier than data packets with lower priorities in the case of medium access contention. A natural consequence is that the control delay of a high-priority loop is usually shorter than that of a low-priority loop. There is no doubt that longer delays make control performance worse. Furthermore, for given delays, a control loop with better performance could intuitively be impacted less significantly than a control loop with worse performance. Based on this observation, we assign higher priorities to control loops that currently have worse performance so that the impact of delays is alleviated.

Of course, the ultimate goal of priority modification is to improve the overall control performance. The basic rule used to assign priorities is that the worse the current performance of a control loop the higher the priority it will be assigned. As a prerequisite for priority re-assignment, a proper performance metric J_i' should be determined. Similar to the choice of J_i, there are many different feasible forms for J_i'. In this chapter we define J_i' as follows based on Eq.(7.5):

$$J_i'(j) = w_i J_i(j) = w_i |e_i(j)|. \tag{7.8}$$

Using this performance index, the rule for determining the priorities can be stated as: the bigger the value of J_i', the higher the relevant control loop's priority. Intuitively, for control loops with the same J_i', the order of their priority levels will not be changed.

Since J_i varies over time, the priorities of control loops might be modified frequently due to even small changes in perturbations in some control loops. To reduce the number of unnecessary switches of priorities caused by small variations of J_i' values and to avoid too serious fluctuations [JIN05], we introduce the notion of *priority switch threshold* δ. With this notion, priority switches are permitted only when the absolute difference between the corresponding J_i' values is no less than δ. Accordingly, we have the following rules for priority modification, where m and n are indexes of control loops.

Rule 1 If $\max\{J_i'\} - \min\{J_i'\} < \delta$, then make no change of all priority levels.

Rule 2 If $J_m'(j) = J_n'(j)$, then maintain the order of current priority levels.

Rule 3 If $J_m'(j) > J_n'(j)$ and $p_m(j-1) > p_n(j-1)$, assuming there is not any control loop whose J_i' value is between $J_m'(j)$ and $J_n'(j)$ (hereafter the same), then $p_m(j) > p_n(j)$.

Rule 4 If $J_m'(j) > J_n'(j)$, $p_m(j-1) < p_n(j-1)$ and $J_m'(j) - J_n'(j) \geqslant \delta$, then $p_m(j) > p_n(j)$.

Rule 5 If $J_m'(j) > J_n'(j)$, $p_m(j-1) < p_n(j-1)$ and $J_m'(j) - J_n'(j) < \delta$, then $p_m(j) < p_n(j)$ and $p_\ell(j) < p_m(j)$, where ℓ is the index of the control loop whose J_i' value is the maximum among all those that are smaller than $J_n'(j)$.

The reason for using RULE 5 is to simplify the process of priority reassignment in some particular situations where $J_m'(j) > J_n'(j) > J_\ell'(j)$ and $p_m(j-1) < p_n(j-1) < p_\ell(j-1)$. Note that it cannot be guaranteed that $p_m(j) = p_m(j-1)$ and $p_n(j) = p_n(j-1)$ when $J_m'(j) = J_n'(j)$. We only maintain the order of their priority levels. The values of $p_m(j)$ and $p_n(j)$ may consequently change with the relative J_i' value of each control loop.

In terms of feedback scheduling, the above priority modification method can be regarded as some kinds of direct feedback scheduling scheme, since it determines the priorities of control loops *directly* according to their current performance. Of course, this method is not actually a scheduling algorithm from the viewpoint of real-time scheduling theory. This is because we do not realize any real-time scheduler for operating systems. While the sensors' priorities are modified periodically, the network is still scheduled based on fixed-priority scheduling algorithm at runtime. On the other hand, since no new low-level scheduler is needed for our approach, it is easy to implement in practice.

7.3.4 Design Considerations

Based on the description above, Fig.7.3 gives the pseudo code for the integrated feedback scheduling algorithm.

```
//ρ: Deadline miss ratio
//e: Control error (in control loop )
//h: Sampling period
//p: Loop/sensor priority
Integrated Feedback Scheduling {
    Input: ρ, {e}
    Period Adjustment {
        //Calculate new total utilization U
        Compute ERR using Eq.(7.3);
        Compute U using Eq.(7.2);
        //Reassign sampling periods
        FOR each control loop
            Compute performance index J using Eq.(7.5);
        END
        FOR each control loop
            Compute new sampling period h using Eq.(7.6) or Eq.(7.7);
        END
    }
    Priority Modification {
        FOR each control loop
            Compute J' using Eq.(7.8);
        END
        Sort control loops with decreasing J' values;
        Determine new priorities for sensors according to Rule 1–5;
    }
    Output: {h,p}
}
```

Fig. 7.3 Pseudo code for integrated feedback scheduling

In the following of this section, we will discuss the design of some critical parameters in this algorithm.

Since the feedback scheduler is time-triggered, a proper invocation interval T_{FS} should be determined. It is intuitive that as T_{FS} becomes smaller, the feedback scheduler will become more sensitive to variations in available resource, which benefits the improvement of feedback scheduling performance. However, the decrease of T_{FS} will cause the corresponding computing and communication overheads to increase. On the other hand, to achieve accurate enough measurements of deadline miss ratio, T_{FS} should not be too small. In practice, when choosing T_{FS}, it is often necessary to take into account a set of characteristics of the system, e.g. the magnitudes of control loops' sampling periods and the (estimated) frequency of workload variations, and accordingly make a trade-off between feedback scheduling overhead and sensitivity.

The setpoint for deadline miss ratio ρ_r highly relates to the control loops' robustness to packet losses. Recall that packet losses are only one subclass of deadline misses. Theoretically, one could first use relevant control theory (e.g. [AZI03, LIN03, ZHA01]) to obtain the maximum allowable packet loss ratio $\rho_{i,\max}$, and then choose a proper ρ_r in the range of (0, $\min\{\rho_{i,\max}\}$). Generally, it is not desirable to choose a ρ_r value too close to $\min\{\rho_{i,\max}\}$. If so, runtime deadline miss ratio might exceed $\min\{\rho_{i,\max}\}$ sometimes because of overshoot that occurs in transient processes of the feedback scheduling system. This may possibly jeopardize the stability of some control loops. Therefore, to maintain system stability, it is necessary to preserve enough margin for dynamic variations of deadline miss ratio. It should be noted that no guarantee of system stability can be made explicitly even if a ρ_r smaller than $\min\{\rho_{i,\max}\}$ is selected. In a multi-loop context, there may possibly exist such a control loop, say loop i, that becomes unstable because its actual packet loss ratio is bigger than $\rho_{i,\max}$, whereas the deadline miss ratio of the whole system is still less than $\min\{\rho_{i,\max}\}$ at the same time. Fortunately, to minimize the impact of deadline misses in steady states on control performance, most often we will specify a relatively small ρ_r, which makes it easy to provide stability guarantees.

Appropriate control parameters K_P and K_I must be chosen in Eq.(7.2). Ideally, if the relationship between deadline miss ratio and utilization can be described with mathematical equations, the well-established feedback controller design methods such as pole placement could then be used to obtain these two parameters explicitly. However, due to the inherent complexity of network communications, their relationship is unable to be formulated analytically in most circumstances, particularly when the NCS operates in dynamic environments with uncertainty. For this reason, we tune K_P and K_I by means of simulation experiments in this chapter.

The issue on how to choose the maximum allowable sampling periods $h_{\max}$ has been discussed in Chapter 2, and thereby is skipped here.

In the module of priority modification, the value of the switch threshold δ affects the feedback scheduling performance to some degree. If it is too large, then the advantages of priority modification will be negated. If it is too small, however, frequent switches of priorities cannot be avoided effectively. When choosing δ, the runtime control performance of the system should often be taken into account. It could often be determined based on the magnitudes of J_i' values of control loops.

7.4 Performance Evaluation

In this section we assess the performance of the above-proposed integrated feedback scheduling scheme via simulation experiments. Suppose the control network is of CAN type with a data rate of 25 Kbps. Notice that the data rate of real CAN buses may be much higher. However, it could be considered equivalently that all other bandwidth has been used by other communication entities. For simplicity, all physical processes in the control loops are assumed to be DC motors. State feedback controllers are designed using the pole placement method, realized digitally by means of direct discrete-time design. The relevant process model and desired closed loop poles are given as follows:

DC motor

$$\dot{x} = \begin{bmatrix} -1 & 0 \\ 1 & 0 \end{bmatrix} x + \begin{bmatrix} 1 \\ 0 \end{bmatrix} u, \quad y = [0,1] x. \tag{7.9}$$

Desired closed loop poles (on z-plane): $0.8 \pm 0.3\mathrm{i}$.

The sizes of sampled data packets in all control loops are 10 bytes. The corresponding data packet transmission time is $(10 \times 8)/25 = 3.2$ ms. Weighting coefficients $w_i = 1$. Some parameters of the feedback scheduler are given in Table 7.2.

Table 7.2 Simulation parameters for integrated feedback scheduling

Variable	T_{FS}(ms)	ρ_r(%)	K_P	K_I	h_{max}(ms)	ε	δ
Value	500	5	0.3	0.8	20	0.2	0.2

To evaluate the performance of integrated feedback scheduling under dynamic environments, we simulate different scenarios with both light load and heavy load, and compare the results with traditional NCS design method (denoted Non-FS) that employs fixed sampling periods and the RM scheduling algorithm.

7.4.1 Scenario I: Underload

First consider the cases where the network workload is light. In this context there are only two control loops in the system. It is set that $h_1 = 10$ ms and $h_2 = 12$ ms. The initial requested network utilization is $3.2/10 + 3.2/12 = 0.587$. It can be concluded from Theorem 7.1 that the system is schedulable. By default, Loop 1 holds the high priority. During runtime the inputs to the control loops are square waves with periods of 4 s and 2 s respectively.

Fig.7.4 depicts the responses of control loops under traditional open-loop scheduling. Because the priority of Loop 1 is relatively high, its performance is better than that of Loop 2. Regardless, both control loops achieve satisfactory control performance.

When the integrated feedbacks scheduler is utilized, the resulting system responses are shown in Fig.7.5. It is clear that the performance of both loops is pretty good. Comparing Fig.7.5 with Fig.7.4, it can be found out that the convergence of both control loops becomes quicker under integrated feedback scheduling. The performance improvement of Loop 2 is particularly noticeable.

Fig. 7.6 gives the Integral of Absolute Error (IAE) of each control loop. Compared to the Non-FS case, under IFS the control cost of Loop 1 decreases 14.1%, and that of Loop 2 decreases 50.5%. The total control cost of the system (i.e. the sum of IAE of all control loops) is shown in Fig.7.7. The total control cost

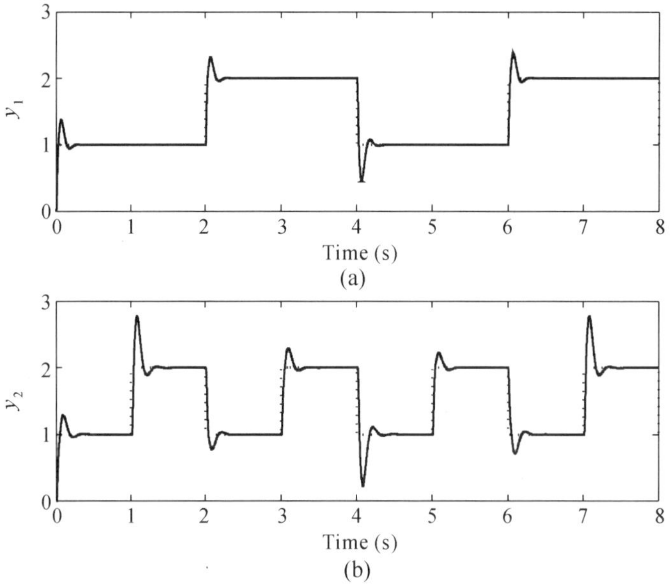

Fig. 7.4 System responses under Non-FS in Scenario I
(a) DC Motor 1; (b) DC Motor 2;

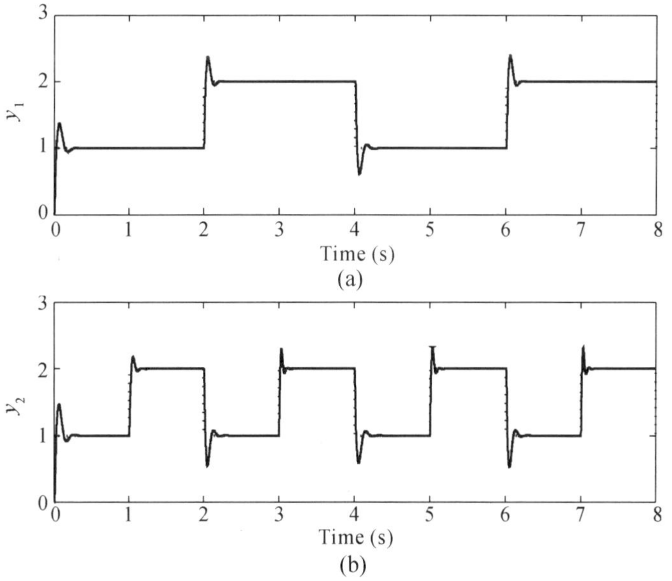

Fig. 7.5 System responses under IFS in Scenario I
(a) DC Motor 1; (b) DC Motor 2;

decreases 40.9% under IFS. Clearly, the QoC is much higher under integrated feedback scheduling than under the traditional design method.

As shown in Fig.7.8, both the sampling periods and the priorities of control

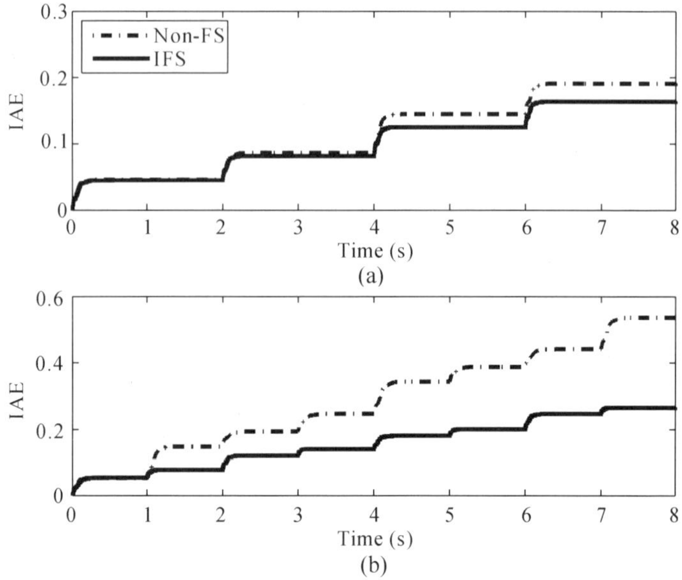

Fig. 7. 6 Integral of Absolute Error of each control loop in Scenario I (a) DC Motor 1; (b) DC Motor 2;

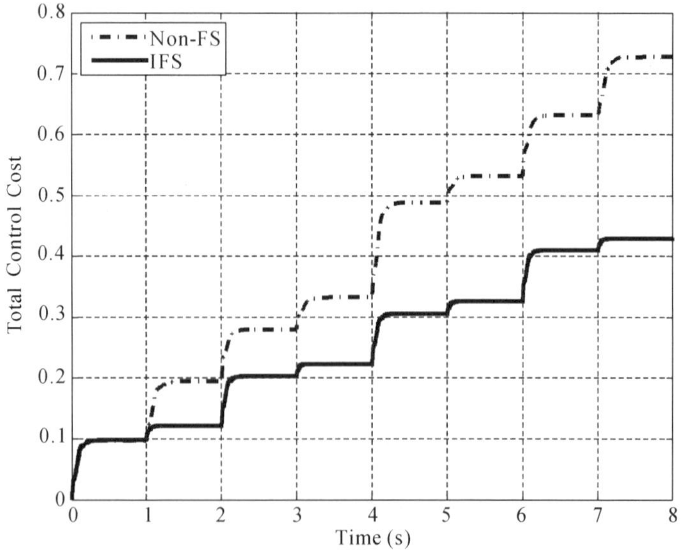

Fig. 7. 7 Total control cost in Scenario I

loops remain fixed at runtime when the traditional design method is used. As a result, the allocation of network resource is fixed, see the upper part of Fig.7.8. Notice that a smaller integer represents a higher priority level in all figures for priorities presented in this chapter.

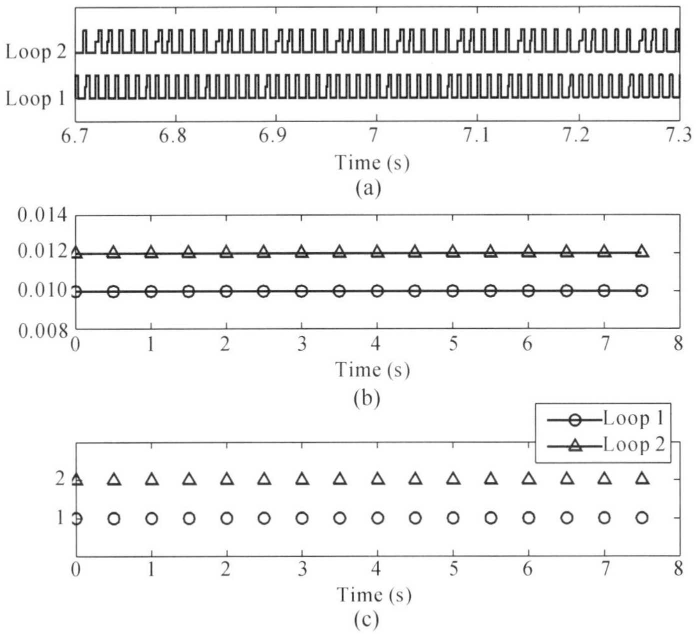

Fig. 7.8 Network schedule, sampling periods and priorities under Non-FS in Scenario I
(a) Network Schedule; (b) Sampling Periods; (c) Priorities

When the integrated feedback scheduler is used, the improvement of control performance mainly benefits from the dynamic adjustment of sampling periods and priorities. As can be seen from Fig.7.9, under integrated feedback scheduling the sampling periods (and the priorities) change at runtime. When the two control loops have comparable performance (in terms of absolute control error), both of the sampling periods will be shortened to make full use of available bandwidth. For instance, at time $t = 6.5$ s, both control loops are in steady states, and consequently their sampling periods are 7.5 ms, which is smaller than both of their initial values (i.e. 10 ms and 12 ms).

When the control performance of two loops is rather different, the feedback scheduler will assign a smaller sampling period and higher priority to the control loop with worse performance. For instance, at time $t = 7$ s, Loop 2 is experiencing a transient process while Loop 1 is steady. In order to bring Loop 2 back to the steady state as soon as possible, it is assigned a relatively small sampling period of 4.3 ms and the high priority with a value of 1. At the same time, the sampling period of Loop 1 is enlarged to 20 ms so that the system schedulability is not violated. In fact, the dynamic changes in sampling periods under integrated feedback scheduling can also be observed from the figure for network schedule. According to the discussions in Chapter 6, enlarging the sampling period of a control loop appropriately when the absolute control error is relatively small will not yield significant degradation of control performance. Consequently, our integrated feedback scheduling scheme results in improvement of overall control performance.

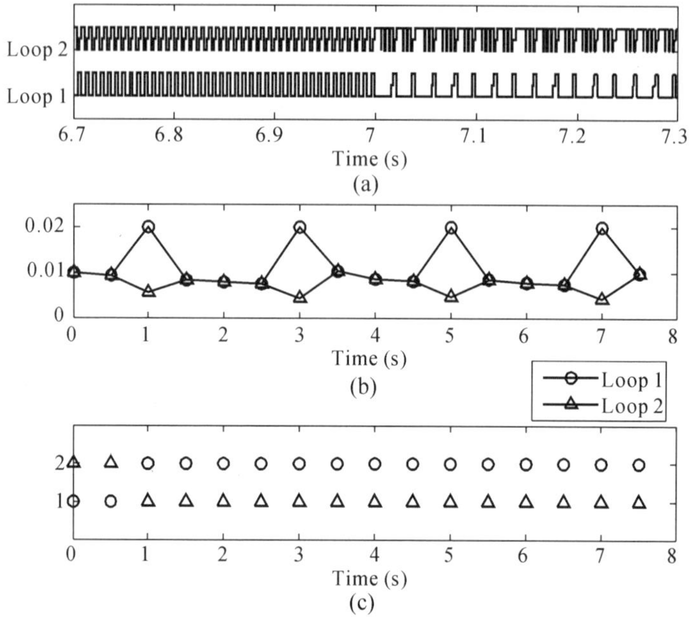

Fig. 7.9 Network schedule, sampling periods and priorities under IFS in Scenario I
(a) Network Schedule; (b) Sampling Periods; (c) Priorities

When the traditional design method is employed, the total requested network utilization of control loops remains at a relatively low level of 58.7% during runtime, see Fig.7.10. In this case, a lot of network resource is wasted, that is, the network remains idle in a significant amount of time. On the contrary, the requested network utilization under IFS increases gradually approaching around 80%, which implies a higher network utilization than Non-FS. As shown in Fig.7.10, although deadline misses occur when the integrated feedback scheduler is used, the deadline miss ratio remains considerably small (can even be 0) most of the time, without jeopardizing the overall control performance.

7.4.2 Scenario II: Overload

Consider next a networked control system with heavy workload, where there are four control loops with $h_1 = h_2 = 10$ ms and $h_3 = h_4 = 12$ ms. This scenario can be viewed as a consequence of adding two extra control loops onto the system in Scenario I for the purpose of e.g. system update. In this context the initial requested network utilization is $2 \times (3.2/10 + 3.2/12) \times 100\% = 117.4\%$. It is apparent that the system is overloaded and hence unschedulable. With traditional open-loop scheduling methods, deadline misses cannot be avoided. By default, the priorities are set as: $p_1 > p_2 > p_3 > p_4$. The inputs to Loop 1 and Loop 2 are square waves with a period of 4 s, and Loop 3 and Loop 4 have the same square wave input with a period of 2 s.

Fig.7.11 shows the response of each control loop under traditional open-loop

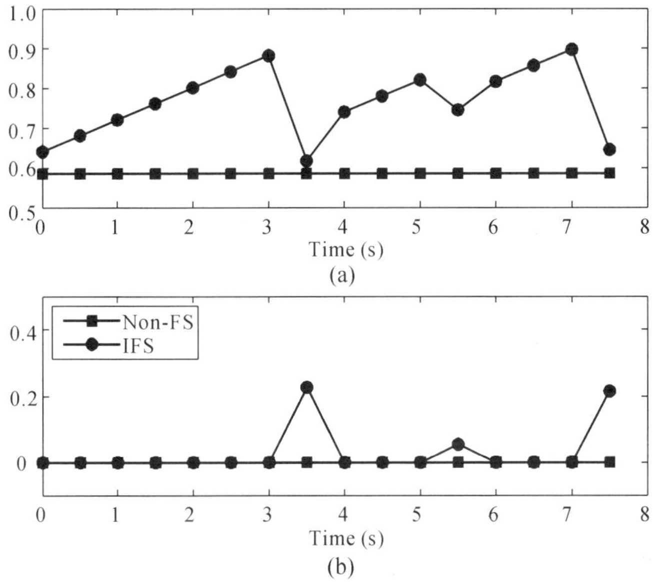

Fig. 7. 10 Requested network utilization and deadline miss ratio in Scenario I
(a) Requested Network Utilization; (b) Deadline Miss Ratio

scheduling. It is clear that Loop 4 finally becomes unstable. Some reasons for this result are argued as follows. On the one hand, this loop has the lowest priority. Consequently its probability of missing deadlines is the highest among all control loops. On the other hand, the requested network utilization is always as high as 117.4%, which is much higher than the schedulable upper bound. This causes too many deadline misses in Loop 4. Since the other three control loops have higher priorities and obtain sufficient network resource, their control performance is satisfactory.

Under integrated feedback scheduling, the system is stable all the time. Furthermore, all control loops achieve pretty good control performance, as shown in Fig.7.12.

Fig.7.13 exhibits the difference between the performance of the examined two schemes in a more straightforward way. For the system in overload conditions, the total control cost under Non-FS increases rapidly because Loop 4 becomes unstable. On the contrary, the total control cost of the system with integrated feedback scheduling increases slowly, indicating relatively good overall performance.

In addition, compared to the case of Non-FS, the QoC of all control loops under IFS is improved, see Fig.7.14. The first three loops' control costs decrease 15.2%, 19.3%, and 20.6%, respectively, and Loop 4 achieves good control performance comparable to that of Loop 3. The instability that occurs under Non-FS is avoided.

Fig.7.15 indicates that under Non-FS Loop 4 has to wait for medium access most of the time because of its lowest priority, which inevitably causes too many deadline misses. This does not happen again when the integrated feedback scheduler is introduced, see Fig.7.16.

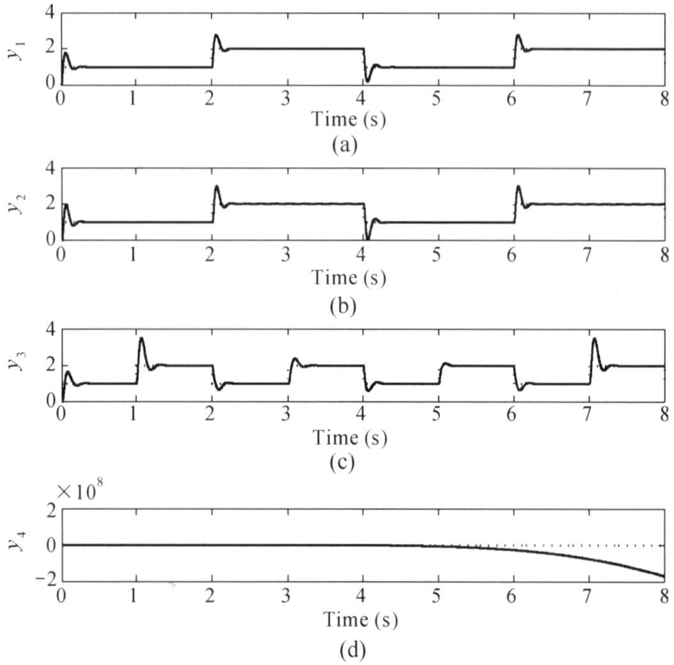

Fig. 7. 11 System responses under Non-FS in Scenario II
(a) DC Motor 1; (b) DC Motor 2; (c) DC Motor 3; (d) DC Motor 4

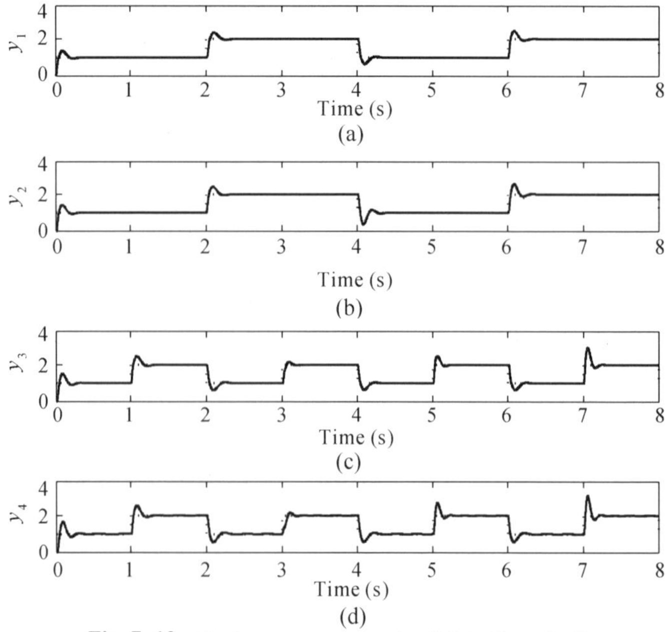

Fig. 7. 12 System responses under IFS in Scenario II
(a) DC Motor 1; (b) DC Motor 2; (c) DC Motor 3; (d) DC Motor 4

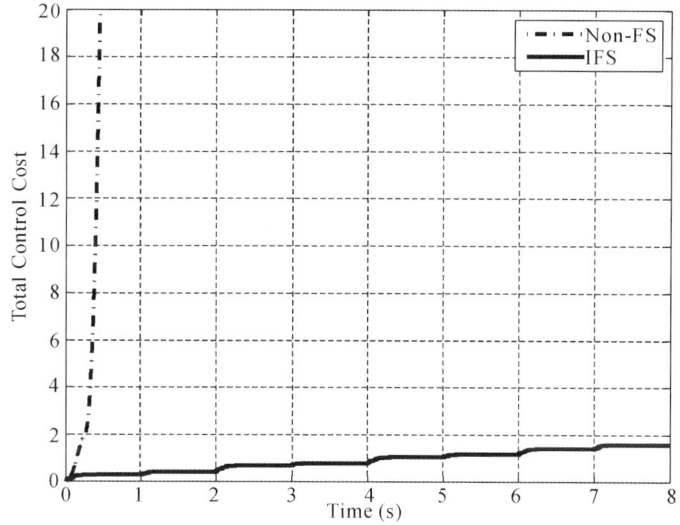

Fig. 7.13 Total control cost in Scenario II

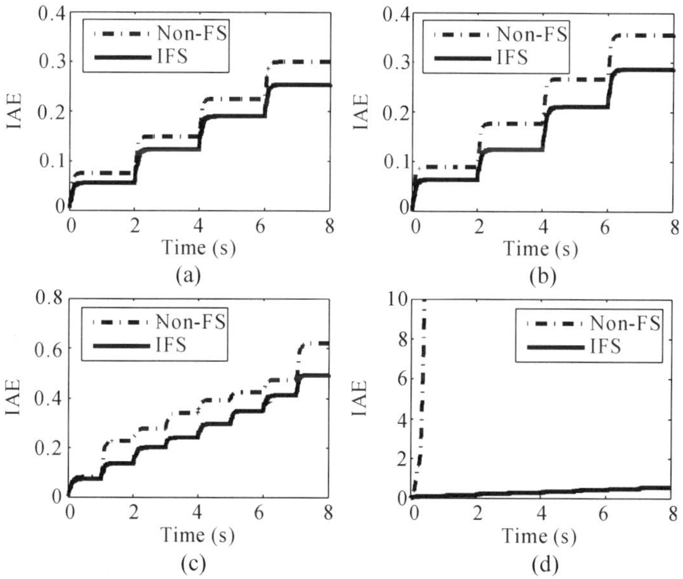

Fig. 7.14 Integral of Absolute Error for each loop in Scenario II
(a) DC Motor 1; (b) DC Motor 2; (c) DC Motor 3; (d) DC Motor 4

In the case of heavy workload, the integrated feedback scheduler prevents the system from continuous unschedulability by adapting the sampling periods of control loops. As can be seen from Fig.7.17, when the performance of control loops is comparable, the feedback scheduler will decrease the requested network utilization through properly enlarging sampling periods, thus decreasing the deadline miss ratio of the system (see Fig.7.19).

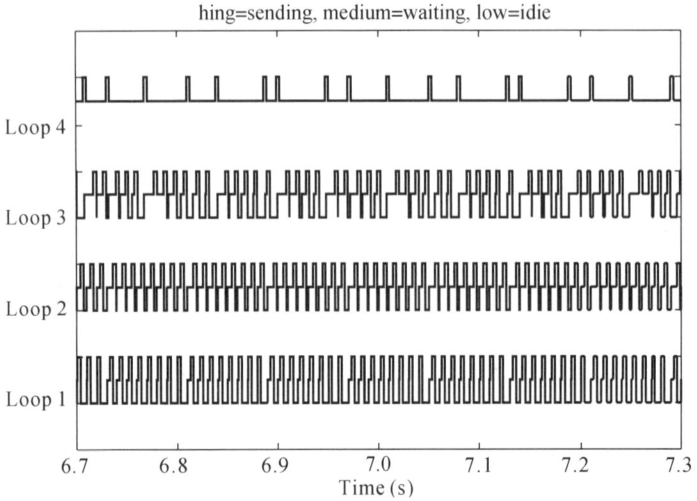

Fig. 7. 15 Network schedule under Non-FS in Scenario II

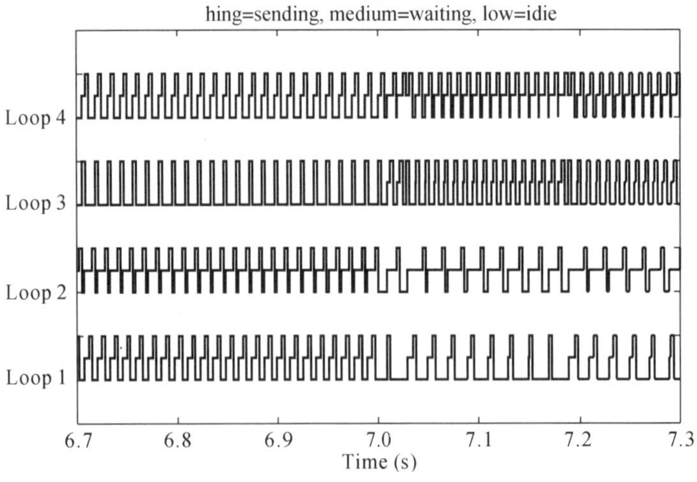

Fig. 7. 16 Network schedule under IFS in Scenario II

When the performance of the control loops is significantly different, on the other hand, the feedback scheduler will assign relatively small sampling periods to the control loops with worse performance, and relatively large ones to the control loops with better performance. Besides sampling periods, the priorities of the control loops will also vary under integrated feedback scheduling, as demonstrated in Fig.7.18. In contrast, both sampling periods and priorities of all control loops are fixed with the traditional design method.

Fig.7.19 presents the requested network utilizations (i.e. network workloads) and deadline miss ratios under different schemes. When the traditional design method is employed, the network workload remains at 117.4% all the time, and the

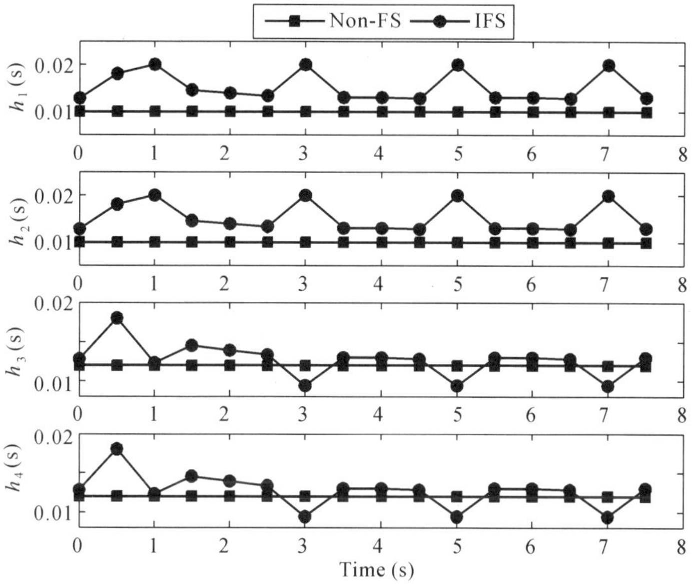

Fig. 7.17 Sampling periods of control loops in Scenario II

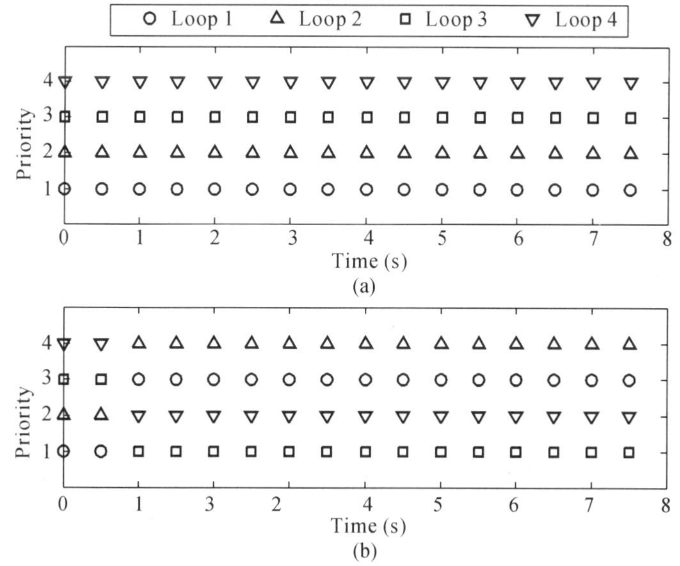

Fig. 7.18 Priorities of control loops in Scenario II
(a) Non-FS; (b) IFS

deadline miss ratio is around 25.5%. In the case of integrated feedback scheduling, the requested network utilization is successfully kept below (and close to) 100%, and the deadline miss ratio approaches a steady state close to 0.

As mentioned above, in order to decrease the deadline miss ratio, the integrated feedback scheduler enlarges the average sampling period of each loop to some degree.

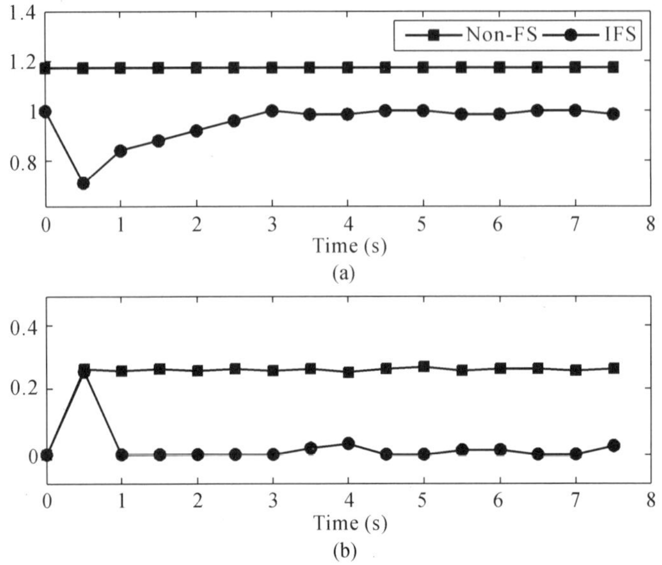

Fig. 7.19 Requested network utilization and deadline miss ratio in Scenario II
(a) Requested Network Utilization; (b) Deadline Miss Ratio

Although in theory larger sampling periods will yield worse control performance, the total control cost of the system decreases (see Fig.7.13) because the deadline miss ratio is decreased. Like in Scenario I, thanks to the dynamic adjustment of sampling periods and priorities, which enables each control loop to obtain as much bandwidth as possible when it needs most, the performance of all control loops is improved (see Fig.7.14).

7.5 Summary

Exploiting the newly emerging methodology of codesign of feedback control and network scheduling, this chapter has presented an integrated feedback scheduling scheme. This scheme combines a cascaded feedback scheduling algorithm for sampling period adjustment and a direct feedback scheduling algorithm for priority modification. It optimizes the use of available network resource by means of dynamic adaptation of both sampling periods and priorities, thus enabling flexible QoC management in networked environments. Using integrated feedback scheduling, the deadline miss ratio of networked control systems could be controlled effectively. The available network resource will be fully used even in underloaded conditions, while graceful degradation of QoC will be achieved in overload conditions. Simulation results argue that the proposed integrated feedback scheduling scheme is able to effectively tackle the problem of bandwidth limitation and workload variation, thus providing a new approach for NCS design and implementation in dynamic environments.

It is not hard to find out that the integrated feedback scheduling scheme facilitates coordinations between different architectural layers, that is, the adjustment of sampling periods and the modification of priorities. According to the protocol stack of CAN, sampling periods are parameters at the application layer, while priority modification is associated with the MAC layer. The proposed integrated feedback scheduling scheme enables information exchange across different protocol layers. From this insight, integrated feedback scheduling can be regarded as some kind of cross-layer design method. We will go further on this topic in the next chapter.

An interesting open issue is the practical implementation of (integrated) feedback schedulers in networked environments. In embedded control systems, see Chapters 3 and 4, a feedback scheduler may be implemented as a task that runs in parallel with control tasks. This makes it much simpler to implement. In distributed control systems, however, the implementation of a feedback scheduler will become difficult. In a networked environment, it is the nodes that act as the real information processing units, but they are spatially distributed. This makes it difficult for the feedback scheduler to e.g. gather information. For NCSs over CAN, the broadcast mechanism of CAN could be used to realize communications between the feedback scheduler and various nodes, while the modification of priorities can be realized by changing the identifier field of CAN data frame. Regardless, it is still non-trivial to examine the extra communication overhead introduced by feedback scheduling.

References

[ALH05] A. Al-Hammouri, V. Liberatore. Optimization Congestion Control for Networked Control Systems. *IEEE INFOCOM' 05 Student Workshop*, 2005.

[AZI03] B. Azimi-Sadjadi. Stability of Networked Control Systems in the Presence of Packet Losses. *Proc. of IEEE Conf. on Decision and Control*, pp.676 – 681, 2003.

[BEN06] M.M.B. Gaid. Optimal Scheduling and Control for Distributed Real-Time Systems. Ph.D Thesis, COSI Laboratory, ESIEE Paris, 2006.

[BOS91] BOSCH, CAN Specification (Version 2.0), 1991.

[BRA02] M.S. Branicky, S.M. Philips, W. Zhang. Scheduling and Feedback Codesign for Networked Control Systems. *Proc. 41st IEEE CDC*, Las Vegas, pp.1211 – 1217, 2002.

[CAS06] R. Castañé, P. Martí, M. Velasco, A. Cervin, D. Henriksson. Resource Management for Control Tasks Based on the Transient Dynamics of Closed Loop Systems. *18th Euromicro Conf. on Real-Time Systems*, Dresden, Germany, 2006.

[EKE00] J. Eker, P. Hagander, K.E. Årzén. A Feedback Scheduler for Real-Time Controller Tasks. *Control Engineering Practice*, Vol. 8, No. 12, pp.1369 – 1378, 2000.

[HE04] J. He, H. Zhang, Y. Jing. An Integrated Control and Scheduling Optimization Method of Networked Control Systems. *Journal of Electronic Science and Technology of China*, No. 02, pp.57 – 60, 2004.

[HEN05] D. Henriksson, A. Cervin. Optimal Online Sampling Period Assignment for Real-Time Control Tasks Based on Plant State Information. *Proc. of the 44th IEEE Conf. on Decision and Control and European Control Conf. ECC*, Seville, Spain, 2005.

[JIN05] H. Jin, H.A. Wang, H. Wang, D. Wang. Scheduling Control Tasks with Threshold-based Largest Dedication First. *International Journal of Information Technology*, Vol. 11, No. 5, pp.61 – 70, 2005.

[LI05] S.B. Li. Study on Robust Control Algorithm of Networked Control Systems. Ph.D Thesis, Zhejiang University, 2005.

[LI06] S.B. Li, Y.Q. Wang, F. Xia, Y.X. Sun. Guaranteed Cost Control of Networked Control Systems with Time-Delays and Packet Losses. *International Journal of Wavelets*, Multi-resolution and Information Processing, Vol. 4, No. 4, pp.691 – 706, 2006.

[LIN03] Q. Ling, M.D. Lemmon. Soft Real-Time Scheduling of Networked Control Systems with Dropouts Governed by a Markov Chain. *American Control Conf.*, Denver, Colorado, 2003.

[MAR04] P. Martí, C. Lin, S. Brandt, M. Velasco, J.M. Fuertes. Optimal State Feedback Based Resource Allocation for Resource-Constrained Control Tasks. *25th IEEE Real-Time Systems Symposium* (RTSS04), Lisbon, Portugal, 2004.

[SET96] D. Seto, J.P. Lehoczky, L. Sha, K.G. Shin. On Task Schedulability in Real-Time Control Systems. *Proc. 17th IEEE RTSS*, Washington, DC, pp.13 – 21, 1996.

[SHA90] L. Sha, R. Rajkumar, J. Lehoczky. Priority Inheritance Protocols: An Approach to Real-Time Synchronization. *IEEE Trans. Computers*, Vol. 39, No.9, pp.1175 – 1185, 1990.

[TIP03] Y. Tipsuwan, M.Y. Chow. Control Methodologies in Networked Control Systems. *Control Engineering Practice*, Vol. 11, pp.1099 – 1111, 2003.

[VEL04] M. Velasco, J.M. Fuertes, C. Lin, P. Martí, S. Brandt. A Control Approach to Bandwidth Management in Networked Control Systems. *30th Annual Conf. of the IEEE Industrial Electronics Society (IECON04)*, Busan, Korea, 2004.

[WAL01] G.C. Walsh, H. Ye. Scheduling of Networked Control Systems. *IEEE Control Systems Magazine*, Vol. 21, No. 1, pp.57 – 65, 2001.

[XIA04a] F. Xia, Z. Wang, Y.X. Sun. Design and Evaluation of Event-Driven Networked Real-Time Control Systems with IEC Function Blocks. *Proc. IEEE SMC*, Hague, Netherlands, Vol. 6, pp.5148 - 5153, 2004.

[XIA04b] F. Xia, Z. Wang, Y.X. Sun. Simulation Based Performance Analysis of Networked Control Systems with Resource Constraints. *Proc. of 30th IEEE IECON*, Busan, Korea, 2004.

[XIA04c] F. Xia, Z. Wang, Y.X. Sun. Improving Quality of Control of Networked Control Systems. *Proc. 7th Int. Conf. on Information Technology*, Hyderabad, India, 2004

[XIA05a] F. Xia, X.H. Dai, Z. Wang, Y.X. Sun. Feedback Based Network Scheduling of Networked Control Systems. *Proc. 5th IEEE Int. Conf. on Control and Automation*, Budapest, Hungary, pp.1231 - 1236, 2005.

[XIA05b] F. Xia, H.X. Yin, Z. Wang, Y.X. Sun. Theory and Practice of Real-Time Scheduling in Networked Control Systems. *Proc. 17th Chinese Control and Decision Conf.*, Harbin, China, pp.1251 - 1256, 2005.

[XIA06a] F. Xia, X.H. Dai, Y.X. Sun, J.X. Shou. Control Oriented Direct Feedback Scheduling. *International Journal of Information Technology*, Vol. 12, No. 3, pp.21 - 32, 2006.

[XIA06b] F. Xia, L.P. Liu, S.B. Li, Y.X. Sun. Integrated Feedback Scheduling of Networked Control Systems. *Dynamics of Continuous Discrete and Impulsive Systems-Series B: Applications and Algorithms*, *Special Issue on ICSCA'06*, Watam Press, pp.3274 - 3280, 2006.

[XIA07a] F. Xia, Y.X. Sun. Cascaded Feedback Scheduling of Multi-Loop Networked Control Systems. *Chinese Journal of Information and Control*, Vol. 36, No. 3, pp.328 - 333, 2007.

[XIA07b] F. Xia, Y.X. Sun, Y.C. Tian. Feedback Scheduling of Priority-Driven Control Networks. Submitted, 2007.

[YEP03] J. Yépez, P. Martí, J.M. Fuertes. The Large Error First (LEF) Scheduling Policy for Real-Time Control Systems. *Work-in-Progress Session of the 24th IEEE Real-Time Systems Symposium (RTSS03)*, Cancun, Mexico, 2003.

[ZHA01] W. Zhang. Stability Analysis of Networked Control Systems. Ph.D Thesis, Department of Electrical Engineering and Computer Science, Case Western Reserve University, 2001.

[ZHA06a] W.H. Zhao, F. Xia. A Neural Network Approach to QoS Management in Networked Control Systems over Ethernet. *Lecture Notes in Control and Information Sciences*, Vol. 344, pp.444 - 449, 2006.

[ZHA06b] W.H. Zhao, F. Xia. Design and Simulation of Function Block Based Networked Control Systems. *Proc. 1st IEEE ICICIC'06*, Beijing, China, Vol. 3, pp.287 - 290, 2006.

Chapter 8

Cross-Layer Adaptive Feedback Scheduling

There is a strong trend towards wireless networked Real-Time Control Systems. However, a variety of factors such as the variability of wireless channel capacities, multi-path fading, and ambient interference make it extraordinarily difficult to design and implement Wireless Control Systems (WCSs). To attack the uncertainty in available communication resource in WCSs, this chapter presents an Adaptive Feedback Scheduling scheme that takes advantage of the methodology of cross-layer design. The proposed approach exchanges information across the physical layer and the application layer in the presence of noise interference and link transmission rate variations. The overall QoC is guaranteed through controlling the deadline miss ratio of control loops. Design parameters of the feedback scheduler are adapted to dynamic changes in wireless channel capacity. To further improve the feedback scheduling performance, an event-driven invocation mechanism for feedback schedulers is suggested. Simulation results demonstrate that this approach is efficient to deal with channel capacity variations and noise interference, thus providing an enabling technology for wireless control in dynamic environments. It is also argued that in this context event-driven invocation mechanism improves the performance of the feedback scheduling system.

8.1 Introduction

In recent years there are witnessing rapid evolution and successful applications of wireless technologies in many fields such as consumer electronics. As a newly emerging research topic in the area of real-time control, Wireless Control Systems (WCSs) are attracting ever-increasing attention from both academia and industry. In a WCS, spatially distributed nodes such as sensors, controllers and actuators are connected with wireless links, which facilitate data transmission and information exchange.

The use of wireless networks has obvious advantages [MAT05, WIL05] as compared to wired networked control systems that have become widespread today (see Chapter 7). For instance, various difficulties related to the installation and maintenance of the large number of cables normally required are completely gotten rid of, thus the flexibility and expandability of the system can be further enhanced. As a result, system maintenance and update become easier, and, of course, the cost will be cut down. In some harsh industrial environments it is forbidden or unfavourable to use cables due to constraints concerning, e.g. physical environments and production conditions. This is especially the case when deleterious chemicals, severe vibrations and high temperatures are present that could potentially damage any sort of cabling. For such situations wireless technologies offer a much better choice for achieving connectivity. In addition, control over wireless networks satisfies the requirements of mobile systems, thus making it possible to realize closed loop control of mobile objectives such as automated guided vehicles, mobile robots, and unmanned aerial vehicles.

A few efforts have been made on applying wireless technologies such as Bluetooth (IEEE 802.15.1), WLAN (IEEE 802.11, also called WiFi), and ZigBee (IEEE 802.15.4) to control systems, for example, [RAM05]. This area is in its infancy at this moment, with most of work done by academic institutions. Regardless, some commercial companies have launched projects on developing various related products for real-world applications. For instance, Ember designed a mesh network for control of water treatment processes [TUC03]; ABB developed the so-called *wireless proximity sensor* that offers a solution making significant progresses in addressing reliability and energy conservation issues for fully wireless closed loop control systems [APN03]. Meanwhile, the deployment costs of wireless systems are decreasing continuously. For example, for a typical process control system comprised of 200 devices, installing wired fieldbus is estimated to cost \$70, 000, while the installation costs for Bluetooth, WLAN and ZigBee are approximately \$34, 000, \$17, 500 and \$12, 000, respectively [MAT05]. Relatively cheaper installation costs make wireless technologies more attractive for industrial control applications.

8.1.1 Motivation

Despite obvious benefits, the use of wireless networks raises new challenges for control systems design. Wireless channels have adverse properties such as path loss, multi-path fading, adjacent channel interference, Doppler shifts, and half-duplex operations [WIL05]. Consequently, transmitting radio signals over wireless channels could inevitably be affected by many factors, such as ambient noise, physical obstacles, node movement, environmental changes and transmit power, just to mention a few. In contrast to traditional wired control networks that usually have fixed communication capacity, the link capacity of wireless channels may vary over time. Due to its significant non-determinism, communication systems over wireless channels are far less dependable than wired networked systems [ARZ06].

From a control point of view, the use of communication networks gives rise to problems related to delay, data loss, and jitter. Compared to the case of wired networked control, these problems become more pronounced in control systems closed over wireless networks. In general, the bit error rate of a wireless channel is several times that of a wired connection, which indicates that the probability of data packets being delayed or even lost due to more retransmissions would increase significantly [PLO04]. Because the operations of wireless transceivers are half-duplex, wireless systems cannot support non-destructive MAC protocols with random medium access. This might possibly cause communication delay and its jitter to increase due to contention and waste of the communication resource. There is no doubt that these phenomena degrade the QoC, even causing system instability in extreme circumstances. Therefore, in designing control systems over wireless networks, it is of paramount importance to effectively address temporal problems such as data loss, delay and jitter induced by communications over wireless channels.

While the area of wireless control systems is board with a foreseeable bright prospect, in which both the control and communication communities are involved, much less work has been done to date in this area as compared to the field of wired networked control. A few existing methods for traditional networked control systems have been extended or modified to facilitate wireless control. From the perspective of real-time communication, some researchers, [YE01] for example, attempt to provide control applications with dependable communications by modifying established communication protocols for wireless networks. Some other researchers, mainly from the control side, instead try to design controllers robust to the temporal non-determinism of wireless networks based on theoretical or experimental analysis of delay and packet loss associated with specific communication protocols, for example, [NIK05, SCH06].

An alternative approach that seems more efficient is codesign of control and wireless communication, especially for systems with limited communication resource, such as a control system where multiple sensors and actuators share the same wireless channel. Partially because of its interdisciplinary nature, this codesign approach is intuitively complicated and hard, with very few results reported. Liu and Goldsmith [LIU04] introduce cross-layer design methodology into WCS design, and present a four-layer framework, but they do not consider online adaptation of the sampling periods of control loops. In [KAW05, PLO04] the authors propose to adjust dynamically the sampling periods, but do not deal with the variability of wireless channel capacity. Their algorithms use fixed parameters without adaptation. In addition, to the best of our knowledge, applying feedback scheduling techniques to WCSs is nearly unexplored.

8.1.2 Contributions

This chapter is about the design of WCSs closed over IEEE 802.11b WLANs[1], taking into account the sharing of a single wireless channel among multiple communication entities. The primary objective is to provide QoC guarantees for

---

1 Unless otherwise specified, WLAN is used hereafter to refer to IEEE 802.11b.

WCSs by means of flexible resource management in dynamic environments that feature (noise) interference and node movement.

To reduce the negative effect of deadline misses on control performance, a simple deadline miss compensation method is used. Based on cross-layer design methodology, an Adaptive Feedback Scheduling (AFS) scheme [XIA07] is presented. This method takes advantage of sharing and exchanging of information across the physical layer and the application layer within the dominating communication protocol stack. The sampling periods of control loops are adjusted dynamically with respect to current channel condition in order to control the deadline miss ratio and ultimately improve the control performance. To cope with the dynamic variations of link capacity, the feedback scheduler uses a simple proportional control algorithm with adaptive parameters. Since radio interference and node movement in wireless systems are stochastic and unpredictable in most situations, it is usually hard to select an appropriate invocation interval for a time-triggered feedback scheduler. To address this problem, an event-driven invocation mechanism for feedback schedulers is suggested. This mechanism contributes not only to reduction of feedback scheduling overheads (on average), but also to quick responses to changes in communication resource availability, thus resulting in further improve-ment of the feedback scheduling performance.

The rest of this chapter is organized as follows: Section 8.2 describes the architecture of the control system considered, followed by a simple deadline miss compensation method. The case study system used as an illustrative example throughout this chapter is also described. In Section 8.3 the employed cross-layer design methodology is illustrated. The temporal properties of the studied wireless control system in terms of deadline miss ratio is analysed by simulation experiments. The Adaptive Feedback Scheduling method is presented, together with relevant algorithms. Section 8.4 focuses on the event-driven invocation mechanism for feedback schedulers. In Section 8.5 the effectiveness of the proposed event-triggered Adaptive Feedback Scheduling approach is validated by simulation experiments highlighting its advantages. Finally, Section 8.6 concludes with discussions on possible extensions over the proposed approach.

8.2 System Model

Consider a WCS shown in Fig.8.1, where, besides an interfering loop, there are altogether N independent control loops. Each control loop consists of a smart sensor (S), a smart actuator (A), a controller (C) and a physical process (P). To facilitate time synchronization, assume that the sensor and the actuator run on top of the same clock platform. The nodes communicate using the IEEE 802.11b protocol. The computation times of all control tasks on the controllers are assumed to be negligible relative to communication delays. The total delay within a control loop is consequently equal to the sum of the communication delay of sampled data (i.e. system

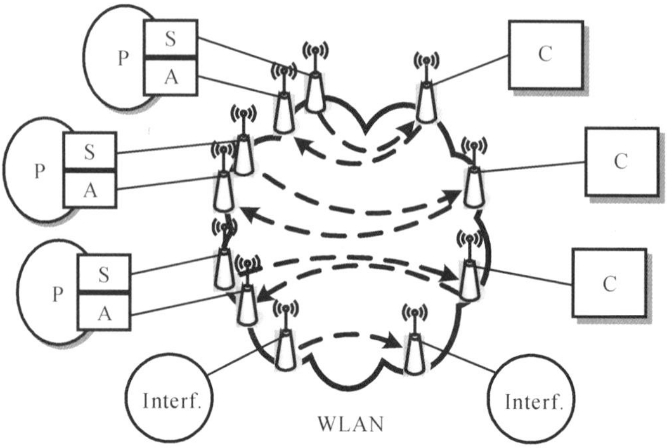

Fig. 8. 1 Architecture of a wireless control system

output) from the sensor to the controller and the communication delay of control command (i.e. control input) from the controller to the actuator, including both waiting delays and transmission delays. The purpose of using smart sensors is to support online adjustment of sampling periods. The smart actuators are used to facilitate handling deadline misses.

An example of the timing diagram of data transmission in one loop of the above system is given in Fig.8.2, where y, u, and h denote the system output, the control command, and the sampling period, respectively. Intuitively, data packets, y_2 and u_6 for example, might possibly be lost during the process of transmission. Another phenomenon worthy of attention is that in a certain sampling interval there may be more than one control commands arriving at the same actuator, which is sometimes called *multiple sampling*. In Fig.8.2, for instance, two control commands, u_3 and u_4, are received in the fourth sampling interval. Since the actuators are event-triggered, multiple sampling may yield irregularity of operation timing of the actuators. Meanwhile, it also implies that there must be no control action in at least one sampling interval, which is often referred to as the problem of *vacant sampling*.

To minimize the impact of these phenomena, it is made a rule that each actuator would act only once in every sampling period. In particular, if the control command associated with current sampling period arrives no later than its deadline, then the actuator performs the control action immediately. Otherwise, an event of deadline miss occurs in this sampling period, and the smart actuator will compensate for it accordingly, using a method given below. Notice that a control command associated with any other sampling period is illegal to trigger the operation of the actuator in a specific sampling period, even when it is the only control command received in this sampling period. For instance, the actuator will not act when u_5 is received in the period h_6.

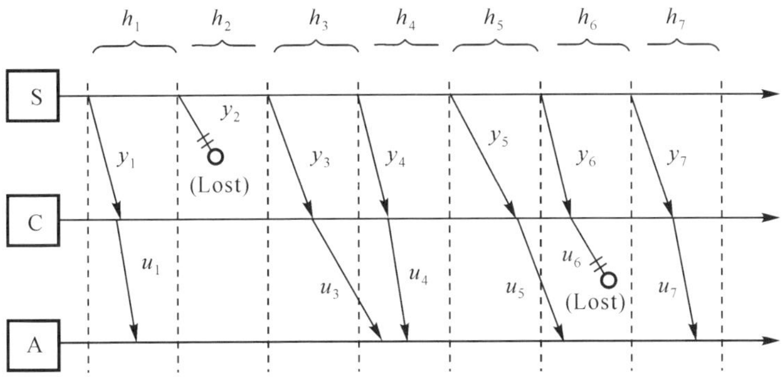

Fig. 8.2 Timing diagram of a wireless control loop

8.2.1 Dealing with Deadline Misses

In the context of wireless control, there are generally two classes of deadline misses. The first is that the sampled data or the control command is truly lost in the transmission process because of factors such as bit errors, noise interference, and/or low received signal strengths. As a consequence, the control command will never arrive at the actuator. This is the case for y_2 and u_6, for example, as shown in Fig.8.2. In contrast, in the second class of deadline misses, the control command is actually received by the actuator, but the communication delay exceeds the deadline, which is equal to the sampling period. Examples of this kind of deadline misses include u_3 and u_5 in Fig.8.2.

Since wireless radio communication is not deterministic in nature, and furthermore, WLAN employs a MAC protocol of random medium access type, most often WCSs suffer from significantly higher and more stochastic deadline miss ratios relative to wired networked control systems. Undoubtedly, too frequent deadline misses jeopardize control performance. Therefore, it is important to compensate for deadline misses in the context of wireless control.

In the literature, many compensation methods for deadline misses have been presented, e.g. [LIN03, SCH06]. While these algorithms have proved effective, they are typically characterized by high complexity, and hence unsuitable for practical applications. Another issue that must be taken into account is the limited computation power of the smart actuators, which cannot accomplish timely computations for overly complex algorithms.

Based on these observations, a simple and practical method is proposed to deal with deadline misses. This method is inspired by the nqPID algorithm in [VAR03], and performs very well when applied to workload prediction in Dynamic Voltage Scaling systems, being insensitive to parameter changes. In [TIA06], Tian, *et al.* use a simplified form of the nqPID algorithm to deal with deadline misses.

If the deadline is missed in the k-th sampling period, i.e. the actuator has not received $u(k)$ by the end of the k-th sampling period (or the beginning of the next sampling period), then the actuator estimates $u(k)$ using Eq.(8.1) and performs the action on the controlled process accordingly before the next sampling operation.

$$\hat{u}(k) = \widetilde{K}_{\mathrm{P}} u(k-1) + \widetilde{K}_{\mathrm{I}} \sum_{i=k-m}^{k-1} \frac{u(i)}{m} + \widetilde{K}_{\mathrm{D}}[u(k-1) - u(k-2)], \quad (8.1)$$

where $\hat{u}(k)$ is the estimate of $u(k)$, $\widetilde{K}_{\mathrm{P}}$, $\widetilde{K}_{\mathrm{I}}$, $\widetilde{K}_{\mathrm{D}}$ and m are user-specified coefficients. In this chapter, $\widetilde{K}_{\mathrm{P}} = 0.4$, $\widetilde{K}_{\mathrm{I}} = 0.2$, $\widetilde{K}_{\mathrm{D}} = 0.4$, and $m = 5$.

Using Eq.(8.1), the actuator predicts $u(k)$ based on the previous m control inputs. For this purpose, the actuator has to temporarily store these m consecutive control inputs. After $\hat{u}(k)$ is obtained, the actuator stores it and discards $u(k - m)$ at the same time. If $u(k)$ is received later, that is, when a deadline miss of the second kind occurs, the actuator replaces $\hat{u}(k)$ in the memory with this accurate $u(k)$ value. A benefit of this operation is that the system can predict the control input with higher precision in the event of another deadline miss. For example, with the timing shown in Fig.8.2, the actuator will calculate the estimate of u_3 at the end of period h_3. It is this estimated value that is acted on the controlled process and is stored at this moment. When u_3 arrives during the term of h_4, the actuator uses it to replace the estimated value of u_3 in the memory. In order to deal with deadline misses in this way, the system has to mark data packets with time stamps, which make the actuator aware of which sampling period the received control input belongs to.

8.2.2 A Case Study

For simple description, consider a wireless control system that consists of three identical control loops. The process under control is a DC motor modelled in continuous-time form [TIP03]:

$$G(s) = \frac{2029.826}{(s + 26.29)(s + 2.296)}. \quad (8.2)$$

The controllers use the PID control law with a continuous-time form given by Eq.(6.12). The controller parameters are: $\widetilde{K}_{\mathrm{P}} = 0.1701$, $\widetilde{K}_{\mathrm{I}} = 0.378$ and $\widetilde{K}_{\mathrm{D}} = 0$. Digital controllers are designed by discretizing continuous-time controllers. As sampling periods are changed, the controller parameters of digital controllers are updated accordingly.

Due to node movement, the communication distance between the controller and the process (where the sensor and the actuator are attached) may change during runtime. According to the properties of wireless signal transmission, the received signal strengths drop with increasing communication distances. When the Signal-to-Noise Ratio (SNR) of the received signals is below a certain level, IEEE 802.11b will make the trade-off between speed and communication reliability by reducing the transmission rate, for example, from the usual maximum value of 11 to 5.5, 2, or even 1 Mbps [IEEE 99, PLO04]. This inherent feature of 802.11b gives rise to variability of channel capacity, a crucial issue that should be taken into account when designing control systems over WLANs.

Apart from the changes in channel capacity, another problem that needs to be addressed is the effect of noise interference on control performance. In the subsequent sections, our solution for these problems will be presented and validated, while using this case as a background example.

8.3 Cross-Layer Adaptive Feedback Scheduling

To enable wireless control in dynamic environments, the feedback scheduling technology is adopted. It has been shown in previous chapters that feedback scheduling is promising in managing uncertainties in resource availability. This motivates the use of this technology in dynamic management of the variable communication resource in wireless control systems. To cope with the adverse properties of wireless communication, the cross-layer design methodology, a technique that is gaining increasing importance in networking applications, is incorporated with feedback scheduling.

8.3.1 Cross-Layer Design Methodology

The design of wireless networks is often based on a layered network protocol stack. To facilitate standardization and implementation with a modular architecture, the design and operation of different network layers are separated. As shown in Fig.8.3, IEEE 802.11b specifies two layers, i.e. the physical layer and the MAC sub-layer, among the seven-layer OSI reference model[2]. The physical layer specifies the transmission media for communication entities, while the MAC layer decides how the communication channel is shared by multiple communication entities. At the physical layer 802.11b specifies four different levels of transmission rates, i.e. 1, 2, 5.5 and 11 Mbps. At the MAC layer 802.11b exploits CSMA/CA to solve resource contention among multiple nodes. The layered architecture of WLAN simplifies network design, thus contributing a lot to its well-known popularity. In the context of wireless control, it is intuitive that the control applications are at the application layer.

From a real-time communication point of view, the temporal behaviour such as delay and deadline miss ratio of WLAN in data transmission relies heavily on the system configuration at the physical layer, e.g. signal transmit power, transmission rate, etc. Furthermore, diverse factors including environmental changes and node movement may possibly result in different characteristics of the temporal behaviour. Therefore, when the system operates under dynamic environments, the timing properties of WLAN may vary with different physical layer parameters, which influence in turn the performance of the control systems at the application layer.

---

2 The seven layers of OSI reference model from bottom to top are the Physical layer, the Data-Link (MAC) layer, the Network layer, the Transport layer, the Session layer, the Presentation layer, and the Application layer.

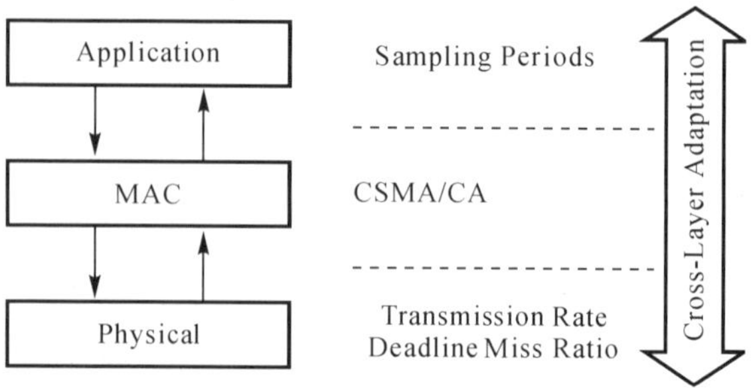

Fig. 8.3 Cross-layer design framework for Wireless Control Systems

To maximize control performance, it is necessary to take advantage of dynamic interaction between the physical layer and the application layer. Cross-layer design should be exploited to achieve application adaptation. In fact, it has been found that making use of information from lower layers in decision-making processes at upper layers is able to improve the overall system performance [ICC03, LIN06, LIU04].

In this chapter the sampling periods of control loops are chosen as the parameters at the application layer. The main reason behind this choice is that the sampling periods influence not only the control performance but also the traffic on the network, which affects the communication delay and the deadline miss ratio. In a sense, the deadline miss ratio can be regarded as an indicator for link quality associated with the physical layer. Since the transmission rate may change at runtime, it naturally becomes another parameter at the physical layer. Consequently, the primary role of feedback scheduling that exploits cross-layer design is to adjust the sampling periods of control systems at the application layer based on relevant information such as deadline miss ratio and transmission rate from the physical layer. As Fig. 8.3 shows, the aim is to optimize the use of resource by means of dynamic interaction between multiple layers, and to ultimately improve the control performance.

In wireless networked systems a straightforward design objective of feedback scheduling is controlling the deadline miss ratio at a desired level. Sine wireless networks cannot support non-destructive communication protocols, it is impossible to analyse the system schedulability for WCSs over WLANs. Traditional real-time scheduling theory, e.g. Theorems 2.1 and 2.3 cannot be applied. As a result, the network utilization is not a suitable choice for the controlled variable.

Without loss of generality, the deadline miss ratio of all control loops is used as the controlled variable of the feedback scheduling system. The deadline miss ratio of interfering signals is not accounted for. Actually, because WLAN employs a MAC protocol featuring random medium access, the deadline miss ratio of control loops also reflects that of the interfering signals.

8.3.2 Analysis of Deadline Misses in WLAN

Before designing the feedback scheduling algorithm, the temporal behaviour of

WLAN in terms of deadline miss ratio needs to be examined. In the following the effects of the transmission rate r and the sampling period h on the deadline miss ratio ρ are analysed by simulation experiments.

Assume that there is no interfering signal in the system shown in Fig.8.1, and the sizes of all data packets to be transmitted over the network are 1 KB. Fig.8.4 depicts the deadline miss ratio of the system with different transmission rates and different sampling periods. For each pair of transmission rate and sampling period, the simulation is run 10 times, with each lasting 3 seconds. Each data value given in Fig.8.4 is the arithmetical mean of the deadline miss ratios recorded separately in these 10 runs.

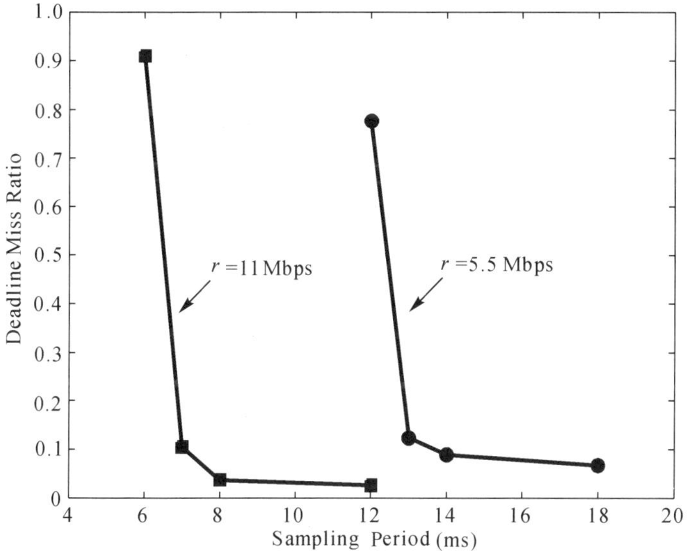

Fig. 8.4 Deadline miss ratio as a function of sampling period and transmission rate

Examine first the effect of sampling period on deadline miss ratio. Since the three control loops in the system are completely identical, every change in the sampling period shown in Fig.8.4 implies that the sampling periods of all three control loops are adjusted at the same time. With a given transmission rate, the deadline miss ratio decreases as the sampling period increases, and the rate of the change also decrease gradually.

When the sampling period is relatively small, the network traffic is heavy, resulting in frequent collisions among communication nodes. As a result, the communication delay of a data packet may become large, even go beyond the deadline, or the data packet may be dropped out due to too many retransmission attempts. In such situations, the deadline miss ratio will naturally be large. Enlarging the sampling period reduces the deadline miss ratio. The reasons behind can be explained as follows. First, the increase of sampling period reduces the amount of network traffic and hence the probability of node collisions. Consequently, the communication delay of data packets decreases on average, and the possibility of data packets being dropped also decreases thanks to the reduction in the number of retransmission attempts. Second, as the sampling period increases, the deadlines of data packets to be transmitted increase accordingly. Intuitively, larger deadlines allow longer communication delays, which reduce the possibility of data packets missing their deadlines. There is no doubt that both of these effects result in reduced deadline miss ratios.

Although increasing sampling periods reduces the deadline miss ratio, which improves control performance, the established sampled-data control theory says that the increase of sampling period may adversely degrade the QoC, even jeopardizing system stability. Therefore, a compromise between the deadline miss ratio and the sampling period must be made to achieve the best possible control performance.

Examine next the effect of the transmission rate r on the deadline miss ratio. Comparing the results for $r = 5.5$ Mbps and $r = 11$ Mbps, it can be seen that larger transmission rate benefits the reduction of deadline miss ratio, especially when the communication resource is scarce. For instance, when $h = 12$ ms, the deadline miss ratios for $r = 5.5$ Mbps and $r = 11$ Mbps are $\rho = 77.6\%$ and 2.7%.

Since this chapter primarily considers the changes in wireless channel capacity caused by node movement, the change of WLAN transmission rate implies that the communication distance between relevant nodes, i. e. the sensor/actuator and the controller, has changed. According to the principle of WLAN, the node distance for $r = 5.5$ Mbps must be longer than for $r = 11$ Mbps. As the distance increases, the strengths of received signals become weaker, causing the possibility of packets being lost to increase. Meanwhile, because of the decrease of transmission rate, the probability of nodes colliding with each other becomes larger, which in turn leads to longer communication delays. Therefore, compared with the case of $r = 11$ Mbps, deadlines are more prone to be missed when the transmission rate is 5.5 Mbps. This is the case even if the sampling periods are sufficiently large. For example, when $r = 5.5$ Mbps and $h = 18$ ms, the deadline miss ratio is $\rho = 6.7\%$, which is larger than the deadline miss ratio $\rho = 2.6\%$ for $r = 11$ Mbps and $h = 12$ ms.

8.3.3 Adaptive Feedback Scheduling Algorithm

Based on the above analysis, the basic idea of feedback scheduling of wireless control systems can be stated as follows: *with the goal of achieving optimized control performance, dynamically adjust the sampling periods of control loops to maintain the deadline miss ratio at a desired level*. From the control perspective, lower deadline miss ratios are always better. Therefore, large sampling periods should be used to avoid deadline misses. In Chapter 7, the integrated feedback scheduler is designed taking advantage of this methodology.

However, it is not easy to completely avoid deadline misses in a typical wireless environment. As also shown in Fig.8.4, in order to reduce the deadline miss ratio to a near-zero level, quite a large sampling period has to be assigned to each control loop. Unfortunately, such a large increase in sampling period could degrade the control performance remarkably. In this context, the final control performance of the system may be adversely deteriorated, regardless of the decrease in the deadline miss ratio. Therefore, in wireless control systems it could be favourable to maintain the deadline miss ratio at an appropriate non-zero level.

Within the framework of feedback scheduling, a simple proportional control algorithm is used to adjust the sampling period:

$$\Delta h(j) = K \cdot e(j), \tag{8.3}$$

where K is the proportional coefficient, $e(j) = \rho(j) - \rho_r$ is the difference between actual deadline miss ratio and its desired value, and j is the index of the invocation instant of the feedback scheduler. Alternatively, the control error of deadline miss ratio could also be re-defined here as done in Chapter 7. Taking into account the constraint on the maximum allowable sampling period $h_{\max}$ of the control loops, the sampling period at j-th instant is then calculated by:

$$h(j) = \min\{h(j-1) + K \cdot e(j),\ h_{\max}\}. \tag{8.4}$$

The main reasons why a simple proportional control law is used in the above equations are that it simplifies online computations with a small overhead, and also makes it convenient to adapt parameters.

In the above algorithm, the proportional coefficient K and the deadline miss ratio setpoint ρ_r are two important design parameters. In the following related design considerations are described.

Proportional Coefficient

For simplicity, assume that with a given transmission rate the relationship between the deadline miss ratio and the sampling period can be approximately described by:

$$\rho = f(h). \tag{8.5}$$

In the vicinity of the operating point $h = h_r$ and $\rho = \rho_r$, good performance could be achieved by setting:

$$K = \frac{1}{\left|\frac{\partial f}{\partial h}\right|}. \tag{8.6}$$

Fig.8.4 indicates that $\left|\frac{\partial f}{\partial h}\right|$ is not constant. It decreases significantly as the sampling period increase. Therefore, it is necessary for the feedback scheduling algorithm Eq.(8.3) to adapt the value of K to current network conditions at runtime. The accurate description of the function f is very difficult, if not impossible, to obtain in most cases. This is particularly true when the system runs on top of wireless networks characterized by random medium access. When this function cannot be obtained from theoretical analysis, it may be possible to collect a large amount of data through extensive simulations and/or experiments and to fit the curve with a feasible model. For the method presented in this chapter, mathematical description of f is not required. It is used here only for the purpose of demonstrating the requirements of Eq.(8.3) on parameter adaptation. Furthermore, it also helps to make the decision on how to adjust K.

When the deadline miss ratio is at a high level (relative to the desired level), the changes in sampling period have significant effects on the deadline miss ratio, as shown in Fig.8.4. The deadline miss ratio could then be brought back to the desired level by only a small change of the sampling period. Accordingly, the value of K should be set relatively small. Otherwise, when the deadline miss ratio is at a low level, the effect of the change in sampling period on the deadline miss ratio is not so significant. To achieve the desired level of deadline miss ratio more quickly, the value of K should be set larger.

A simple algorithm given by Eq.(8.7) is used in this chapter to adapt the value of K.

$$K = \begin{cases} K_0/2 & \rho(k) > \rho_r + \Delta\rho^+ \\ K_0 & \rho_r - \Delta\rho^- \leqslant \rho(k) \leqslant \rho_r + \Delta\rho^+ , \\ 2K_0 & \rho(k) < \rho_r - \Delta\rho^- \end{cases} \tag{8.7}$$

where K_0 can be obtained from simulation experiments, $\Delta\rho^+$ and $\Delta\rho^-$ are user-specified parameters. It is apparent that the value of K_0 will depend on f and h_r if $K_0 = \frac{1}{\left|\frac{\partial f}{\partial h}\right|_{h=h_r}}$. That is, K_0 will change with the transmission rate r and the deadline miss ratio setpoint ρ_r.

Besides the above equation, there are of course a lot of advanced algorithms that could potentially be more efficient in adjusting the value of K, for example, the gain scheduling method from adaptive control theory. However, these complex algorithms also add online computations associated with the feedback scheduler, thus causing larger overheads.

Deadline Miss Ratio Setpoint

Before describing how to adapt the deadline miss ratio setpoint ρ_r, let's illustrate first how to select ρ_r. For a given control system, the effects of sampling period and deadline miss ratio on the control performance is deterministic, while the deadline miss ratio is related to the sampling period in some way, e.g. Eq.(8.5). Therefore, for a given system setup, there exists an optimal operating point, say (h_r, ρ_r), at which the system will in principle achieve the optimal control performance. Ideally, the best feedback scheduling performance could be achieved by setting the desired level of deadline miss ratio to this optimal point. In practice, the relationships between the control performance, the deadline miss ratio, and the sampling period are very complicated, and therefore cannot be explicitly described. Most often a deadline miss ratio setpoint close to the optimal could be chosen by simulations or experiments.

As shown in Fig. 8.4, for the same sampling period, a larger transmission rate generally yields a smaller deadline miss ratio. This indicates that the relationship between the deadline miss ratio and the sampling period differs from one transmission rate to another. On the other hand, for a given control system, the location of the optimal operating point depends on the relationship between the deadline miss ratio and the sampling period. When the transmission rate is low, to avoid control performance degradation caused by significant impact of the sampling period, it is desirable to choose a relatively large setpoint for deadline miss ratio.

For the sake of simple description, suppose the point A (h_{r1}, ρ_{r1}) in the schematic diagram Fig.8.5 is the setpoint for $r = 5.5$ Mbps, which is quite close to the optimal operating point. When the transmission rate changes, e.g. from 5.5 to 11 Mbps, the operating point of the system will be B(h_{r2}, ρ_{r1}) if a fixed deadline miss ratio setpoint is used. Since tradeoffs should be made between the deadline miss ratio and the sampling period in order to obtain the optimal operating point, it is not difficult to find out that if A is the optimal operating point for $r = 5.5$ Mbps, then B will not be the optimal operating point for $r = 11$ Mbps. The optimal sampling period for $r = 11$ Mbps will instead be some value, say h_{r3}, that falls in the interval (h_{r2}, h_{r1}). This suggests that when the transmission rate increases, the control performance could be improved through decreasing the value of ρ_r, for example, using ρ_{r2} as the new setpoint.

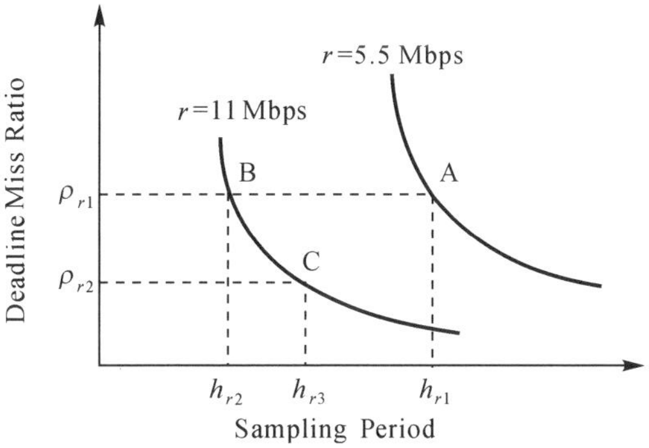

Fig. 8.5 Schematic diagram for adapting deadline miss ratio setpoint

Therefore, we propose to adapt the deadline miss ratio setpoint to different transmission rates. The setpoints used in this chapter are given by:

$$\rho_r = \begin{cases} 10\% & \text{if} \quad r = 5.5 \\ 5\% & \text{if} \quad r = 11 \end{cases}. \tag{8.8}$$

Since only two cases, i.e. $r = 5.5$ Mbps and $r = 11$ Mbps, are considered in simulation experiments, Eq.(8.8) gives only the corresponding two values for ρ_r. Thanks to the fact that the compensation method for deadline misses, which is given by Eq.(8.1), is capable of reducing the negative effects of deadline misses on control performance to some degree, there is actually a considerably large room for choosing the value of ρ_r. Nevertheless, due to the application of the compensation method, it is difficult to formulate analytically the relationship between the control performance and the deadline miss ratio. Therefore, the upper bound of ρ_r that guarantees system stability cannot be obtained explicitly. Regardless, the applicability of the feedback scheduling method will not be affected, while existing results in control theory provide valuable references for choosing ρ_r.

Fig.8.6 presents the pseudo code for the Adaptive Feedback Scheduling algorithm. As canbe seen, this algorithm exhibits online adaptability in two aspects: (1) the adaptation of the proportional coefficient K to deal with the nonlinear relationship between the deadline miss ratio and the sampling period; (2) the adaptation of the deadline miss ratio setpoint to deal with the changes in the transmission rate.

```
Cross-Layer Adaptive Feedback Scheduling {
    //Determine h_max, K_0 at pre-runtime
    //Monitor
    Measure deadline miss ratio ρ;
    Measure transmission rate r ;
    //Adapt parameters if necessary
    IF r changes
        Update K_0
        Update ρ_r using Eq.(8.8)
    END
    Determine K using Eq.(8.7);
    //Compute new sampling periods
    Calculate e ←ρ-ρ_r;
    Calculate Δh using Eq(8.3);
    Reassign sampling periods according to Eq.(8.4);
}
```

Fig. 8. 6 Pseudo code for cross-layer Adaptive Feedback Scheduling

8. 4 Event-Triggered Invocation

Usually, feedback schedulers are time-triggered, that is, they execute as periodic tasks. An obvious advantage with this mode is that it makes convenient to design and analyse the feedback schedulers using feedback control theory and technology, for example, as done in Chapter 5. This is because sampled-data control theory in existence basically originates from periodic sampling. In some situations, however, it could be very difficult to choose a proper invocation interval for time-triggered feedback schedulers.

The efficiency of the time-triggered invocation mechanism could be examined in terms of response speed and overhead. To achieve quick response, feedback schedulers prefer small invocation intervals so that almost all changes in e.g. workload and available resource can be treated in a timely fashion. Since the execution of feedback schedulers consumes resources, which is originally limited, the decrease of invocation interval yields the increase of feedback scheduling overheads, which could adversely influence the system performance.

Compared with the scenarios considered in previous chapters, the environmental changes including noise interference and node movement in wireless networked systems are more uncertain and irregular. This makes it more difficult to apply time-triggered feedback schedulers to wireless control systems. On the one hand, the invocation interval cannot be set very small because accurate deadline miss ratios would otherwise be impossible to obtain. Thus the feedback scheduler will not be capable of coping with, in a timely fashion, interference and node movement that occur between two consecutive invocation instants. On the other hand, it is possible that the system stays in a steady state for quite a long time, when there is actually no need for sampling period adjustment. In this situation, time-triggered feedback schedulers could potentially waste resources in periodic executions of the feedback scheduling algorithm and unnecessary updates of system parameters.

From the above observations, an event-triggered invocation mechanism is presented to improve the efficiency of feedback schedulers in the context of wireless control. Discussed below is how to implement this mechanism.

Design Methodology

The schematic diagram for the event-triggered invocation mechanism is depicted in Fig. 8.7. Similar to the structure of event-based controllers [ARZ99], there are basically two parts in this invocation mechanism [XIA05], the event detector and the feedback scheduling algorithm. The event detector is time-triggered with a period of T_{ED}, while the feedback scheduling algorithm is triggered by the *execution-request* event issued by the event detector. This event-triggered invocation mechanism is applicable not only to the cross-layer Adaptive Feedback Scheduling method proposed in this chapter, but also to most feedback scheduling methods in a general sense.

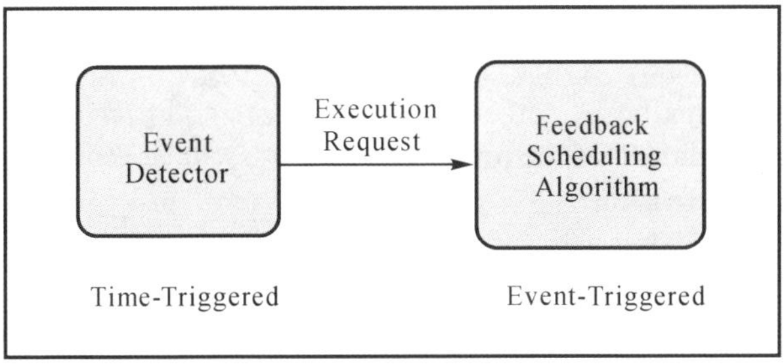

Fig. 8.7 Schematic diagram for event-triggered invocation

The key to implementing the event-triggered invocation mechanism is the design of the event detector. The major role of the event detector is deciding under what conditions the system needs to execute the feedback scheduling algorithm. As mentioned previously, the goal of the feedback scheduling loop is to maintain the deadline miss ratio at a desired level. Intuitively, when the deadline miss ratio is in or close to steady states, there is no need for executing the feedback scheduling algorithm. On the contrary, if the deadline miss ratio has significantly deviated from the desired level, then it becomes mandatory to run the feedback scheduler to adjust system parameters. In this chapter the following condition is used for issuing the *execution-request* event:

$$|\rho(k) - \rho_r| \geqslant \delta . \tag{8.9}$$

According to Eq.(8.9), the feedback scheduling algorithm will be executed if and only if the absolute difference between the actual deadline miss ratio and its desired level is no less than a specific threshold δ. In this way, the disadvantages with time-triggered invocation with respect to response speed and overhead are avoided. Furthermore, the negative effect of measurement noises on the deadline miss ratio may naturally be reduced.

There are two important parameters, T_{ED} and δ, to be set. Generally speaking, choosing these parameters demands careful tradeoffs between quick response and low overhead. Fortunately, this is not as difficult as usually expected. Thanks to the small amount of computations associated with Eq.(8.9), it is possible to assign a small enough period T_{ED} to the event detector to achieve quick response while keeping the feedback scheduling overhead small. The magnitude of measurement noises should be taken into account when choosing the value of δ. The system usually allows the deadline miss ratio to fluctuate around the setpoint with small deviations. Therefore, a δ value that is slightly bigger than the magnitude of measurement noises could be used to reduce the number of executions of the feedback scheduler, which reduces runtime overheads.

8.5 Simulation Experiments

This section conducts simulation experiments on the case given in Section 8.2 to assess the performance of the proposed event-triggered cross-layer Adaptive Feedback Scheduling scheme. Although the results are still preliminary at this stage, some potential advantages of the proposed approach will be demonstrated. Consider the following two scenarios:

(1) Scenario I: the controller and the process are close to each other, WLAN operates at 11 Mbps, there is no interfering signals, $\delta = 0.02$, $K_0 = 0.02$;

(2) Scenario II: the transmission rate drops to 5.5 Mbps, the interfering transmitter sends a data packet of 1 KB to the corresponding receiver every 10 ms, $\delta = 0.04$, $K_0 = 0.006$.

One could notice that different δ values are used in these two scenarios. This is because: (1) the deadline miss ratio setpoints for different transmission rates are different; (2) this makes it convenient to compare the event-triggered and time-triggered feedback scheduling, see Subsection 8.5.2.

Table 8.1 summarizes some parameters used in the simulation experiments, where h_0 denotes the nominal sampling period. It is worth noting that completely identical results cannot be guaranteed for each run of the simulation even with the same system setup. This is a natural consequence of the inherent stochastic feature of communications over WLANs. All of the results given below are those representative ones among many others obtained from a variety of simulation runs.

Table 8.1 Parameters for simulation experiments

Variable	h_0(ms)	h_{max}(ms)	T_{ED}(ms)	$\Delta\rho^+$	$\Delta\rho^-$
Value	15	50	500	0.1	0.08

8.5.1 Experiment I: Adaptive Feedback Scheduling vs. Traditional Design Method

In the first set of simulation experiments, the proposed AFS method and the traditional design method without any feedback schedulers (denoted Non-FS) are compared. Since the three control loops are identical and WLAN adopts a random medium access control mechanism without distinguishing between them, all loops are equivalent in principle. Therefore, only the responses of one control loop will be given.

Fig.8.8 depicts the step responses of Loop (i.e. y_1) under different scenarios when the traditional method is used. It can be seen that the system performs quite well when the transmission rate of WLAN is 11 Mbps. However, under the second scenario, i.e. when the transmission rate drops to 5.5 Mbps with interfering signals, the system finally goes unstable.

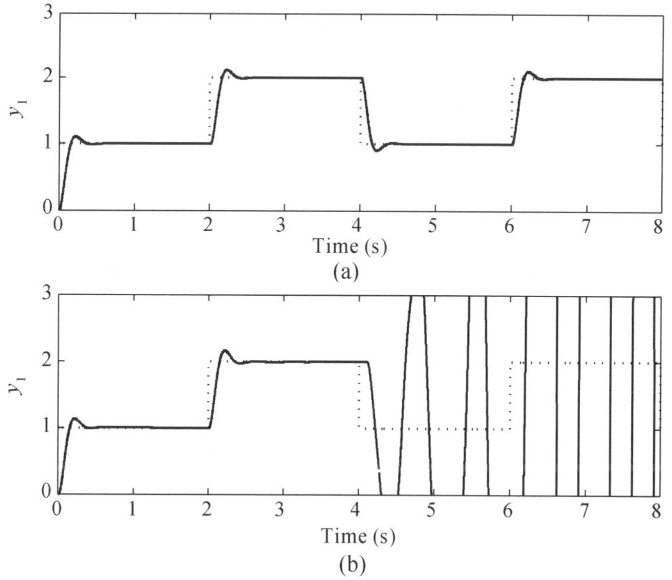

Fig. 8.8 Control performance without feedback scheduling
(a) Scenario I; (b) Scenario II

Fig.8.9 shows the system performance when Adaptive Feedback Scheduling is adopted. The system not only performs well under Scenario I, but also achieves good control performance under Scenario II.

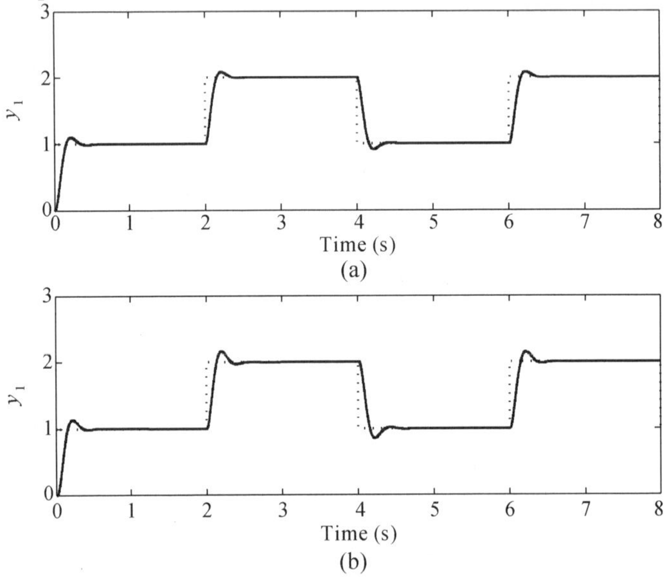

Fig. 8. 9 Control Performance with Adaptive Feedback Scheduling
(a) Scenario I; (b) Scenario II

Fig.8.10 plots the total control cost of the system, which is calculated as the sum of the IAE values of the three control loops. Under Scenario I, the system performance under different schemes is almost identical. Under Scenario II, Adaptive Feedback Scheduling achieves significant improvement of QoC. After time t > 4 s, the total control cost for the traditional method increase quickly, implying that the system goes unstable. With the adaptive feedback scheduler, in contrast, the total control cost remains small all the time. The overall control performance is quite good.

From Fig. 8.10 it can be found out that when the Adaptive Feedback Scheduling scheme is used, the system performs comparably under different scenarios. This can also be seen from Fig.8.9. It can be seen that the proposed Adaptive Feedback Scheduling scheme is able to guarantee high QoC even if the transmission rate of WLAN drops from 11 Mbps to 5.5 Mbps and interfering signals become present.

The sampling periods and the deadline miss ratios under different schemes are shown in Fig.8.11 and Fig.8.12, respectively. With the traditional method, the sampling periods of all control loops are fixed during runtime, i.e. $h \equiv 15$ ms. When WLAN runs on top of 11Mbps (i.e. under Scenario I), the deadline miss ratio is small with a mean of 2.3%. Consequently, the control performance is good. Under Scenario II, the deadline miss ratio remains near 100% after time $t > 4$ s, implying that almost all data packets transmitted on the WLAN miss their deadlines. This inevitably gives rise to system instability.

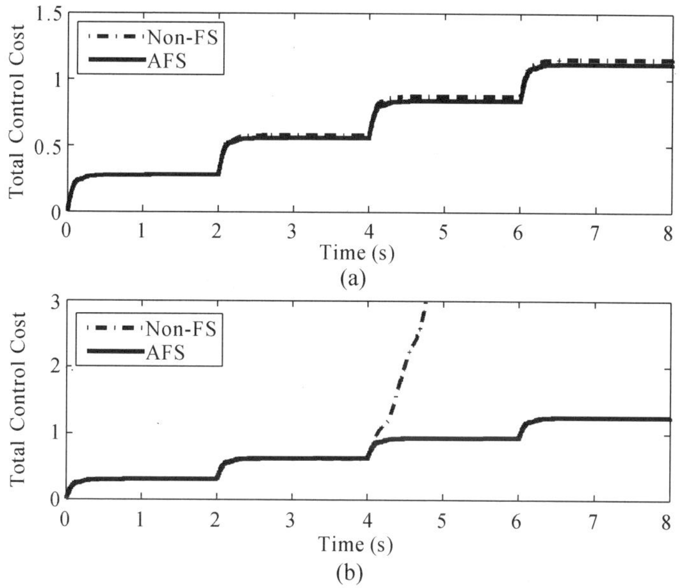

Fig. 8. 10 Total control cost in Experiment I
(a) Scenario I; (b) Scenario II

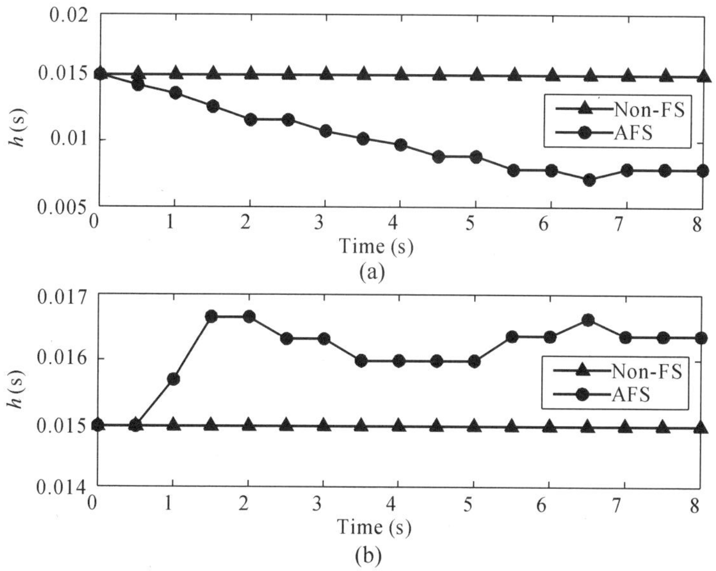

Fig. 8. 11 Sampling periods of control loops
(a) Scenario I; (b) Scenario II

As can be seen from Figs.8.11 and 8. 12, the adaptive feedback scheduler effectively controls the deadline miss ratio through dynamically adjusting the sampling periods. Under Scenario I, the sampling periods of control loops decrease from time

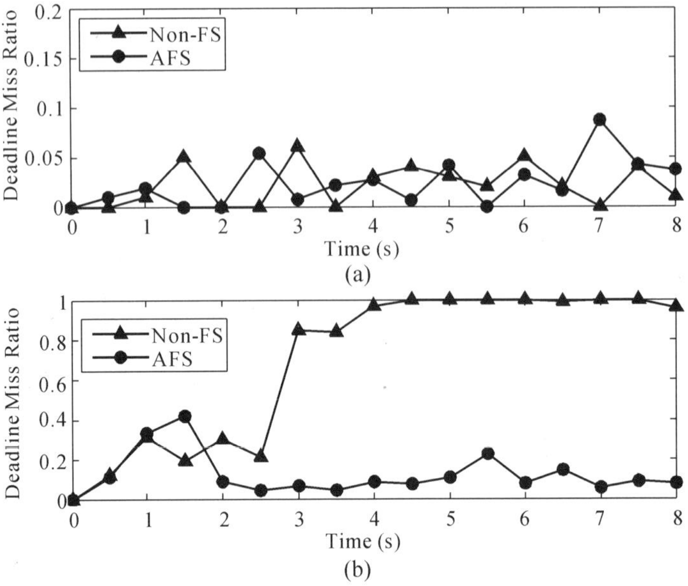

Fig. 8.12 Deadline miss ratio of control loops
(a) Scenario I; (b) Scenario II

$t = 0$, and remain at a nearly steady level, i.e. around 8 ms after time $t = 5.5$ s. During this term, the deadline miss ratio keeps at a low level, with a mean of 2.5%. Finally, it approaches its setpoint 5%. The levels of the deadline miss ratio under both schemes are close, but the sampling periods are smaller with Adaptive Feedback Scheduling. According to sampled-data control theory, the smaller the sampling period the better the control performance. Therefore, Adaptive Feedback Scheduling benefits QoC improvement.

Under Scenario II, the adaptive feedback scheduler successfully avoids overly high deadline miss ratios by increasing the sampling periods gradually. After a transient process, the deadline miss ratio keeps around the setpoint 10%. Thanks to the compensation method used, the control performance is not deteriorated by these deadline misses. One could notice that both the sampling periods and the deadline miss ratio increase on average under Scenario II relative to Scenario I, which may have some negative effects on the control performance. Consequently, the control performance is slightly worse in Scenario II than in Scenario I, as shown in Fig.8.9.

The above simulation results show that for wireless control systems operating in dynamic environments, the Adaptive Feedback Scheduling scheme is able to effectively attack the problem of transmission rate changes and ambient noise interference, thus improving the QoC of the whole system.

8.5.2 Experiment Ⅱ: Event-Triggered vs. Time-Triggered

In the second set of experiments the performance of event-triggered and time-triggered Adaptive Feedback Scheduling methods is compared. To facilitate comparisons with the event-triggered scheme simulated in the first set of experiments, the invocation interval for the time-triggered feedback scheduler is set to $T_{FS} = T_{ED} = 500$ ms.

Now that the results for event-triggered Adaptive Feedback Scheduling have been presented above, the control performance under time-triggered Adaptive Feedback Scheduling will be studied below.

Fig. 8.13 depicts the step response and the Integral of Absolute Error of Loop 1 under Scenario I. Obviously, the control performance in this case is pretty good. Comparing the upper parts of Fig.8.13 with Fig.8.9 it can be seen that under Scenario I the time-triggered and event-triggered Adaptive Feedback Scheduling achieve almost identical control performance.

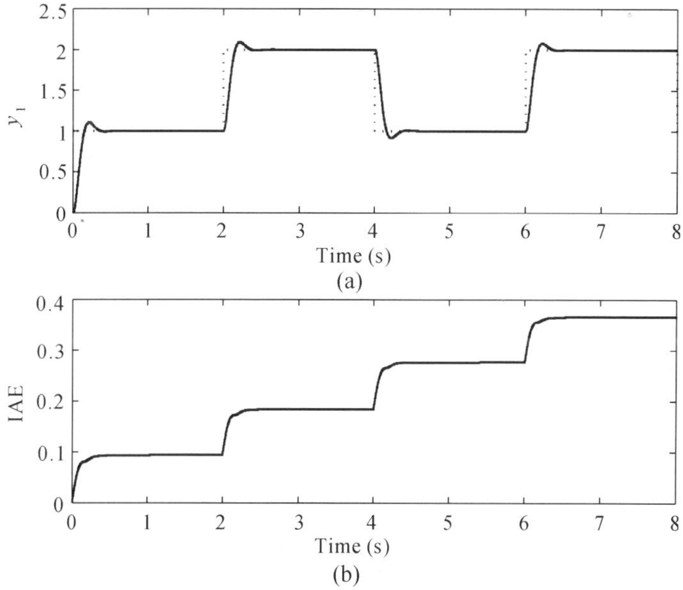

Fig. 8.13 Control performance for time-triggered adaptive feedback scheduling under Scenario I
(a) System Output; (b) IAE

The sampling periods and the deadline miss ratio for the system using time-triggered Adaptive Feedback Scheduling are given in Fig.8.14. It is easy to find out that they vary in like manner as under event-triggered Adaptive Feedback Scheduling. The main difference between them is that with event-triggered feedback scheduling the sampling periods remain unchanged at some consecutive sampling instants, which implies that the feedback scheduler does not actually execute (see the upper part of Fig. 8. 11), whereas the sampling periods are updated at every sampling instant when time-triggered feedback scheduling is used (see the upper part of Fig.8.14).

The response and the IAE of Loop 1 under Scenario II are depicted in Fig.8.15. It is clear that time-triggered Adaptive Feedback Scheduling also delivers good control

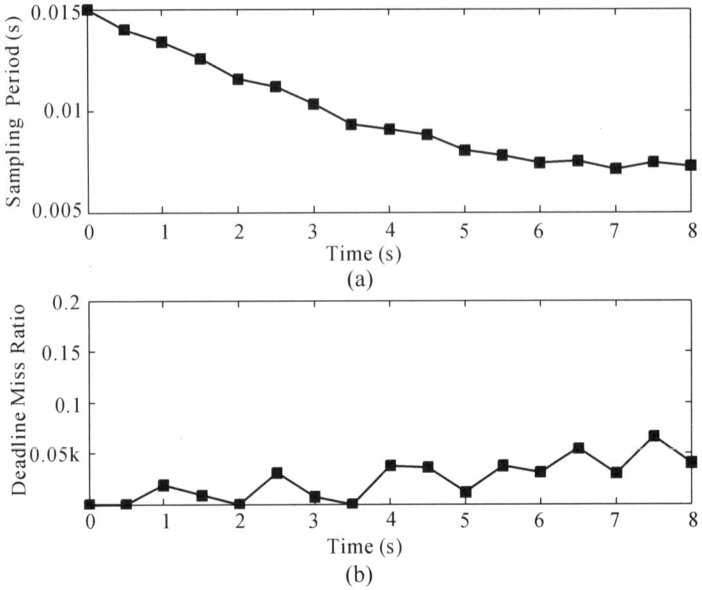

Fig. 8.14 Sampling periods and deadline miss ratio for time-triggered Adaptive Feedback Scheduling under Scenario I
(a) Sampling Period(s); (b) Deadline Miss Ratio

performance. Comparing Fig.8.15 with Fig.8.13 indicates that the control performance becomes slightly worse under Scenario II than Scenario I.

As shown in Fig.8.16, the deadline miss ratio is kept around the setpoint 10% by the feedback scheduler adjusting the sampling periods. However, the sampling periods and the deadline miss ratio are bigger (on average) than in Scenario I, see Figs.8.14 and 8.16. That is why the system performance is better in Scenario I, just as the system performs with event-triggered Adaptive Feedback Scheduling.

Fig.8.17 shows the total control costs of the system with time-triggered and event-triggered Adaptive Feedback Scheduling. An interesting result is that the event-triggered invocation mechanism might possibly result in better control performance than the time-triggered invocation mechanism, though the improvement is not so significant. In Scenario II, for example, the total control cost for event-triggered Adaptive Feedback Scheduling decreases 2.4% in comparison with that for time-triggered Adaptive Feedback Scheduling.

To this point only the control performance is examined, which is actually the ideal system performance without taking into account the effect of feedback scheduler executions. That is, in the above simulation experiments the overhead of feedback scheduling is neglected. For simplicity, the *times of execution* of the feedback scheduler is used as a simple metric for comparing the feedback scheduling overheads.

Table 8.2 summarizes the total control costs and the times of execution of the feedback scheduler with different invocation mechanisms. It is obvious that for different invocation mechanisms the overall QoC remains fairly close. The relative

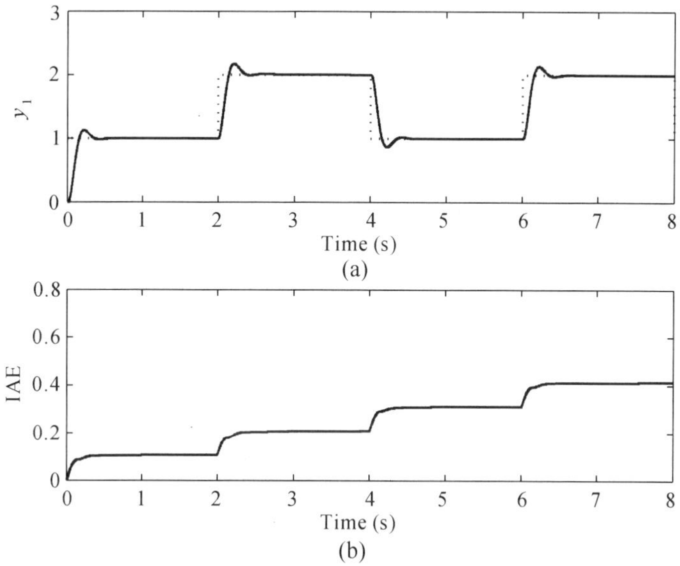

Fig. 8. 15 Control performance for time-triggered adaptive feedback scheduling under Scenario II
(a) System Output; (b) IAE

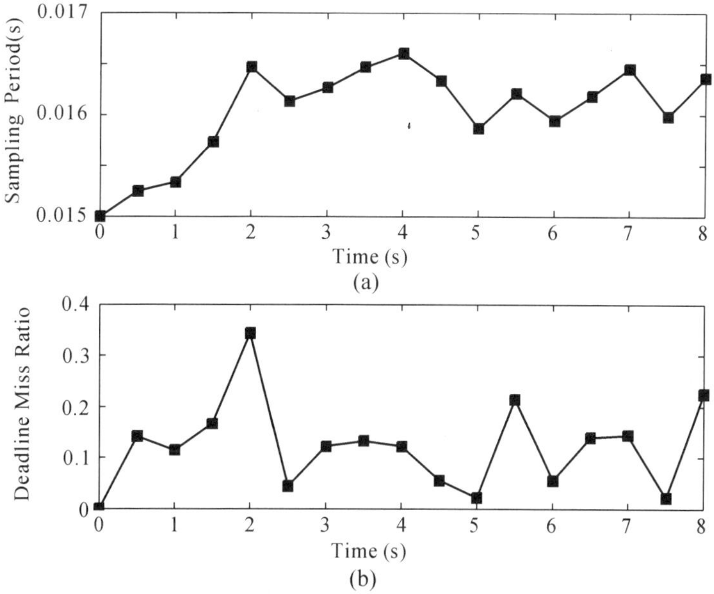

Fig. 8. 16 Sampling periods and deadline miss ratio for time-triggered Adaptive Feedback Scheduling under Scenario II
(a) Sampling Period(s); (b) Deadline Miss Ratio

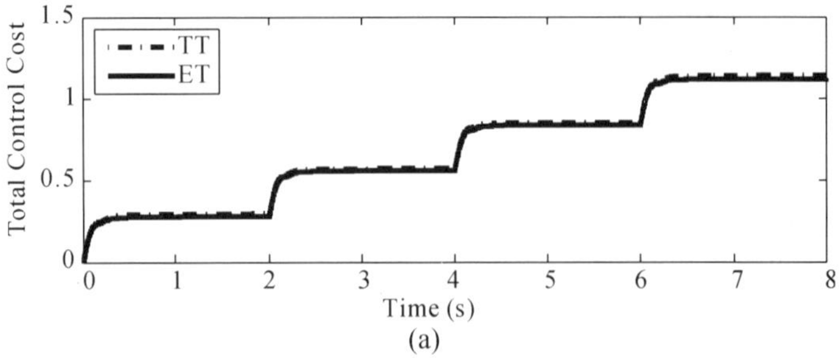

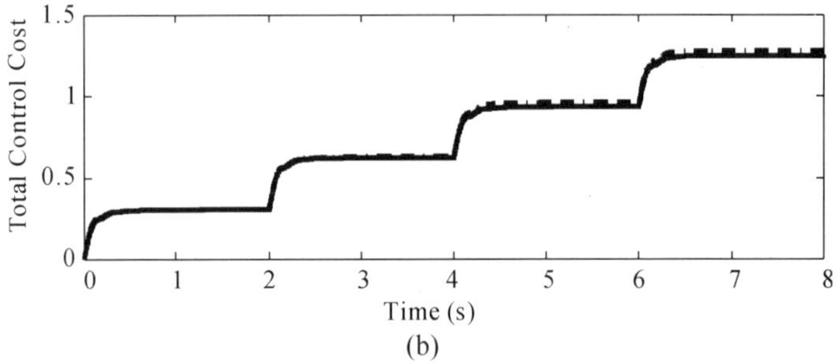

Fig. 8.17 Total control cost in Experiment II
(a) Scenario I; (b) Scenario II

difference of total control costs is less than 3%. In Scenario I, the times of execution of feedback scheduler decreases 31.3% with event-triggered Adaptive Feedback Scheduling as compared to time-triggered scheme. In Scenario II it reduces from 16 to 7, with a relative reduction of 56.3%. The overhead of feedback scheduling is significantly reduced.

Table 8.2 Comparison of event-and time-triggered invocations

	Scenario I		Scenario II	
	TT	ET	TT	ET
$\sum$ IAE	1.148	1.127	1.285	1.254
Times of Execution	16	11	16	7

The above results show that the proposed event-triggered invocation mechanism yields significant reduction in feedback scheduling overhead while providing good feedback scheduling performance, thus improving the efficiency of the feedback scheduling scheme. Furthermore, the event-triggered invocation mechanism can also be used to achieve quicker responses associated with the feedback scheduler without introducing excessively large overheads by simply selecting a smaller T_{ED} value. In this way, the practical performance of feedback schedulers could be further improved.

8.6 Summary

This chapter deals with dynamic management of the communication resource in wireless control systems. Using cross-layer design, an Adaptive Feedback Scheduling scheme has been presented to attack the effects of factors such as noise interference and node movement on control performance. This scheme makes use of related information at the physical layer to adjust the sampling periods of the control systems at the application layer. The control performance is optimized by means of deadline miss ratio control. To meet the requirements deriving from the adverse properties of wireless communications and to avoid the disadvantages of time-triggered invocation in making tradeoffs between response speed and overhead, an event-triggered invocation mechanism has also been suggested, which improves the practical performance of feedback scheduling.

The emerging area of wireless control is currently in its infancy. The method proposed in this chapter offers an effective solution for constructing control systems over wireless networks. However, the present work accomplished by this time is still preliminary. More systematic analysis, both theoretical and experimental, is left for our future work.

The proposed method could be extended in many aspects. One possible extension is improving the cross-layer design framework. For example, to take into account the effect of different communication protocols (i.e. MAC protocols), the MAC sub-layer may be included in the framework of cross-layer feedback scheduling, thus exploring the interaction between the physical layer, the MAC layer, and the application layer. In cases where the energy consumption of the nodes is a concern, physical-layer parameters such as the transmit power may be made available for upper layers.

Another possibility is improving the Adaptive Feedback Scheduling algorithm. Suppose the behaviour of the wireless network in terms of deadline miss ratio could be modelled with sufficient accuracy, for example, an optimized Adaptive Feedback Scheduling algorithm would be easily obtained using the presented design methodology. In this context, even the optimal sampling periods with respect to different system states could possibly be derived analytically using related control theory and technology.

Our future work in this direction also includes the development of an experiment system for wireless control systems, which will be used to evaluate the performance of Adaptive Feedback Scheduling with extensive results.

References

[APN03] C. Apneseth, D. Dzung, S. Kjesbu, G. Scheible, W. Zimmermann. Wireless-Introducing Wireless Proximity Switches. *Sensor Review*, Vol. 23, No. 2, pp.116 – 122, 2003.

[ARZ99] K.E. Årzén. A simple Event Based PID Controller. *Proc. 14th IFAC World Congress*, Beijing, China, Vol. Q, pp.423 – 428, 1999.

[ARZ06] K.E. Årzén, A. Bicchi, S. Hailes, K.H. Johansson, J. Lygeros. On the Design and Control of Wireless Networked Embedded Systems. *Proc. of IEEE Computer Aided Control Systems Design Symposium*, pp.440 – 445, 2006.

[ICC03] ICC panel. Defining Cross-Layer Design in Wireless Networking, http://www.eas.asu.edu/ ~ junshan/ICC03panel.html, 2003.

[IEEE99] International Standard ISO/IEC 8802 – 8811, ANSI/IEEE Std 802.11, Part 11. *Wireless LAN Medium Access Control (MAC) and Physical Layer (PHY) Specifications.* The IEEE, Inc., 1999.

[KAW05] P.A. Kawka, A.G. Alleyne. Stability and Feedback Control of Wireless Network Systems. *Proc. of the American Control Conf. (ACC05)*, Portland, OR, 2005.

[LIN03] Q. Ling, M.D. Lemmon. Optimal Dropout Compensation in Networked Control Systems. *Proc. of the IEEE Conf. on Decision and Control*, 2003.

[LIN06] X. Lin, N.B. Shroff, R. Srikant. A Tutorial on Cross-Layer Optimization in Wireless Networks. *IEEE Journal on Selected Areas in Communications*, Vol. 24, No. 8, pp.1452 – 1463, 2006.

[LIU04] X. Liu, A.J. Goldsmith. Wireless Network Design for Distributed Control. *Proc. IEEE Conf. on Decision and Control*, 2004.

[MAT05] M. Mathiesen, G. Thonet, N. Aakwaag. Wireless Ad-hoc Networks for Industrial Automation: Current Trends and Future Prospects. *Proc. of the IFAC World Congress*, Prague, Czech Republic, 2005.

[NIK05] G. Nikolakopoulos, A. Panousopoulou, A. Tzes, J. Lygeros. Multihoping Induced Gain Scheduling for Wireless Networked Controlled Systems. *Proc. of the 44th IEEE Conf. on Decision and Control*, European Control Conf., Seville, Spain, 2005.

[PLO04] N. Ploplys, P. Kawka, A. Alleyne. Closed Loop Control over Wireless Networks. *IEEE Control Systems Magazine*, pp.58 – 71, 2004.

[RAM05] H. Ramamurthy, D. Lai, B.S. Prabhu, R. Gadh. ReWINS: A Distributed Multi-RF Sensor Control Network for Industrial Automation. *IEEE Wireless Telecommunication Symposium WTS*, Pomona, California, pp.24 – 33, 2005.

[SCH06] L. Schenato. Optimal estimation in Networked Control Systems Subject to Random Delay and Packet Loss. *IEEE Conf. on Decision and Control (CDC 06)*, San Diego, CA, USA, 2006.

[TIA06] Y.C. Tian, D. Levy, M.O. Tadé, T. Gu, C. Fidge. Queuing Packets in Communication Networks for Networked Control Systems. *The 6th World Congress on Intelligent Control and Automation*, Dalian, China, pp.205 – 209, 2006.

[TIP03] Y. Tipsuwan, M.Y. Chow. Control Methodologies in Networked Control Systems. *Control Engineering Practice*, Vol.11, pp.1099 – 1111, 2003.

[TUC03] A. Tuck, C. Burgess. Mesh Meets the Need-Wireless for Industrial Control. ISA Technical Information and Communities, www.isa.org, 2003.

[VAR03] A. Varma, B. Ganesh, M. Sen, S.R. Choudhury, L. Srinivasan, J. Bruce. A control-theoretic approach to dynamic voltage scheduling. *Proc. ACM CASES*, Georgia, USA, pp.255 – 266, 2003.

[WIL05] A. Willig, K. Matheus, A. Wolisz. Wireless Technology in Industrial Networks. *Proceedings of the IEEE*, Vol. 93, No. 6, pp.1130 – 1151, 2005.

[XIA05] F. Xia, Y.X. Sun. Event-Triggered Feedback Scheduling of Embedded Control Systems. *Proc. 1st National Conf. on Pervasive Computing*, Kunming, China, pp.660 – 666, 2005.

[XIA07] F. Xia, Y.C. Tian, C. Peng, Y.X. Sun. Cross-Layer Adaptive Feedback Scheduling of Wireless Control Systems over WLAN. Submitted, 2007.

[YE01] H. Ye, G.C. Walsh, L. Bushnell. Real-Time Mixed Traffic Wireless Networks. *IEEE Transactions on Industrial Electronics*, Vol. 48, No. 5, pp.883 – 890, 2001.

Index